Lecture Notes in Statistics 164

Edited by P. Bickel, P. Diggle, S. Fienberg, K. Krickeberg, I. Olkin, N. Wermuth, and S. Zeger

Springer

New York
Berlin
Heidelberg
Barcelona
Hong Kong
London
Milan
Paris
Singapore
Tokyo

Peter Goos

The Optimal Design of Blocked and Split-Plot Experiments

 Springer

Peter Goos
Department of Applied Economics
Katholieke Universiteit Leuven
Naamsestraat 69
B-3000 Leuven
Belgium
peter.goos@econ.kuleuven.ac.be

Library of Congress Cataloging-in-Publication Data
Goos, Peter.
    The optimal design of blocked and split-plot experiments / Peter Goos.
        p.   cm.— (Lecture notes in statistics ; 164)
    Includes bibliographical references and index.
    ISBN 0-387-95515-1 (alk. paper)
        1. Experimental design.   I. Title.   II. Lecture notes in statistical (Springer-Verlag) ; v.
    164.
    QA249 .G66   2002
    001.4'34—dc21   2002067647

ISBN 0-387-95515-1   Printed on acid-free paper.

Printed in the United States of America.

9 8 7 6 5 4 3 2 1   SPIN 10881903

Typesetting: Pages created by the author using a Springer T$_E$X macro package.

www.springer-ny.com

Springer-Verlag   New York Berlin Heidelberg
A member of BertelsmannSpringer Science+Business Media GmbH

Preface

Quality has become an important source of competitive advantage for the modern company. Therefore, quality control has become one of its key activities. Since the control of existing products and processes only allows moderate quality improvements, the optimal design of new products and processes has become extremely important. This is because the flexibility, which characterizes the design stage, allows the quality to be built in products and processes. In this way, substantial quality improvements can be achieved.

An indispensable technique in the design stage of a product or a process is the statistically designed experiment for investigating the effect of several factors on a quality characteristic. A number of standard experimental designs like, for instance, the factorial designs and the central composite designs have been proposed. Although these designs possess excellent properties, they can seldom be used in practice. One reason is that using standard designs requires a large number of observations and can therefore be expensive or time-consuming. Moreover, standard experimental designs cannot be used when both quantitative and qualitative factors are to be investigated or when the factor levels are subject to one or more constraints. A danger inherent to the use of standard designs is that the experimental situation is adapted to the experimental design available. Of course, it is much better to find the best possible design for the experimental situation at hand. This is exactly the purpose of the approach known as the optimal design of experiments.

In this book, the optimal design approach is applied to two common types of experiments, namely blocked and split-plot experiments. Blocked experiments are needed when not all the experimental observations can be carried out under homogeneous circumstances, for example when more than one batch of material is required or when the experiment takes up more than one day. Split-plot experiments are used when it is impractical to change the levels of some of the experimental factors. A typical example of such factor is temperature because heating a furnace and cooling it down are time-consuming operations. Often, the observations in blocked and split-plot experiments are correlated. This statistical dependence is explicitly taken into account in the optimal design approach as well.

Chapter 1 contains an overview of the experimental design literature. Special attention is given to the basics of the optimal design approach, as well as to the standard response surface designs and to categorical designs. Chapter 2 focuses on the optimal design of experiments with nonhomogeneous error variance and correlated observations. The topic of blocking experiments receives attention as well. In Chapter 3, the design problems considered in this book are described in detail and an appropriate statistical model is introduced. In Chapter 4, optimal designs for blocked experiments are computed. In Chapter 5, the optimal design approach is applied to a blocked optometry experiment. In the Chapters 6, 7 and 8, the optimal design of split-plot experiments is considered. Chapter 9 provides a brief overview of the recent results on two-level factorial and fractional factorial designs. Finally, Chapter 10 summarizes the main results of this book as well as some ideas for future research.

For every type of experiment considered in this book, an algorithm for the computation of optimal designs was developed. A Fortran 77 implementation of the algorithms as well as a number of sample in- and output files can be downloaded from the author's personal website http://www.econ.kuleuven.ac.be/peter.goos/. In case this link fails, the author websites on http://www.springer-ny.com/ will provide an alternative.

While finishing this book, I would like to express my appreciation to those who have contributed to its development and its improvement. Firstly, I would like to thank my advisor Professor M. Vandebroek for her continuing assistance and support during the past few years. I am also indebted to the other members of my doctoral committee, the Professors P. Darius, E. Demeulemeester, A. Donev and W. Gochet, for their invaluable comments and suggestions. I am also grateful to the Fund for Scientific Research–Flanders (Belgium) for providing me with the financial means to carry out the research described in this book, to the Department of Applied Economics of the Katholieke Universiteit Leuven for providing me with a comfortable office and to my colleagues for creating an agreeable working atmosphere. I am also indebted to all people who read through parts of the manuscript and made suggestions and comments, to Scott Chasalow and Kenneth Polse for their help in describing the optometry experiment in Chapter 5, to Steven Gilmour for his help in creating the benchmark design in Chapter 6, to Herlinde Leemans for proofreading the final draft of the book, and to Lieven Tack with whom I have experienced many joyful moments at the office and at international conferences. Finally, I would like to thank my family and friends for their unconditional moral support. In particular, I would like to thank my wife Marijke for her encouragements, patience and understanding.

Peter Goos

Contents

Contents

1
Introduction

Experiments provide an efficient way of learning as long as they are properly designed and analyzed. Because all experimental observations are subject to random error, an efficient design and analysis of experiments requires statistical methods. In this book, we will concentrate on the statistical design of experiments, rather than on their analysis. In the last couple of decades, experimental design has become increasingly popular in quality engineering, but, as will be illustrated in this book, it is used in nearly any field of study: medicine, agriculture, chemistry, etc. As a result, most of the literature is scattered and terminology and methods are very often specific to the area of application.

The purpose of experimenting is to determine and to quantify the relationship between the values of one or more response variables and the settings of a number of experimental variables presumed to affect them. Once this goal is achieved, the experimenter typically tries to identify those settings of the experimental variables that produce the best value of the response variables. The design of the experiment mainly consists of determining the number of experimental runs, the settings of the experimental variables in each of the runs, and the sequence in which the runs have to be executed. For simplicity, it is often assumed that observations are uncorrelated and normally distributed with homogeneous variance. However, this assumption does not hold in many experimental situations. For example, in repeated measurement studies, it is natural to assume that all observations carried out on one subject are correlated. Similarly, the presence of random block effects implies that all runs within the same block are statistically dependent. In other experiments, the assumption of homogeneity is invalid. In cases where the heteroscedasticity is a function of the experimental variables, the behavior of the quality characteristic under investigation can be described by modelling both its mean and its variance. Finally, not all response variables are normally distributed. For instance, the number of defects, a response commonly used in quality environments, usually follows a Poisson distribution. In other cases, the response is a categorical variable or it may represent time to failure. The purpose of this book is to provide useful answers and insights to cope with some of these non-standard experimental situations.

In this book, the focus will be on practical experiments. This implies using discrete or exact designs, rather than continuous or approximate designs. Although the latter often provide useful insights, they only have true practical use when the size of the experiments is large. In that case, the optimal discrete design can be easily obtained from the optimal continuous design. Unfortunately, resource constraints force real-life experiments to be small and other techniques to compute efficient designs need to be used. Although most examples used in this book involve only quantitative factors, the approach presented can handle both quantitative and qualitative experimental factors. Temperature and pressure are typical examples of quantitative variables, whereas machine type and type of material are qualitative variables. Atkinson and Donev (1989) and Cook and Nachtsheim (1989) designed experiments involving both types of variables. Before that time, designing experiments for treatment comparisons, i.e. experiments with qualitative factors, and designing experiments for response surface models or regression models, i.e. experiments with quantitative factors, were two entirely separated specializations.

Standard response surface designs like full and fractional factorial experiments, Plackett-Burman designs, central composite designs or Box-Behnken designs, are popular designs for experiments with quantitative variables only. However, many experimental situations exist where these designs cannot be used. This is due to the fact that standard designs suffer from a lack of flexibility, for instance in cases where the number of observations available to the experimenter is too small to conduct them, in cases where the design region is restricted, or in cases where experimental runs are to be blocked. Moreover, they were developed under the assumption of uncorrelated and normally distributed errors with homogeneous variance. As a result, the attractive properties they possess might not hold if these assumptions are violated. The literature also provides the experimenter with numerous experimental designs for qualitative variables. These categorical designs display shortcomings similar to those of the standard response surface designs. Standard categorical designs are not flexible because they can be constructed only for specific combinations of the number of treatments, the number of blocks and the block size. In addition, their statistical properties are often invalid when the assumption of uncorrelated and normally distributed errors with homogeneous variance no longer holds. In design theory, the development of algorithms for computing optimal designs for a wide range of experimental situations was an answer to the shortcomings of the standard designs.

In this introductory chapter, we provide the reader with a historical overview of experimental design and the three approaches to the statistical design of experiments are highlighted. Firstly, the optimal design theory is presented and a couple of interesting illustrations are given. Secondly, we review the most famous standard response surface designs and, thirdly, we focus on categorical designs. Next, we motivate our choice for the $\mathcal{D}$-optimality criterion to evaluate different design options by examining its advantages with respect to other design criteria. Finally, we review the different approaches to the computation of discrete $\mathcal{D}$-optimal designs when observations are statistically independent and have homogeneous error variance. However, in order to fully understand the purpose of the statistical design of experiments, we start this chapter with a practical design problem and with an introduction to the linear regression model and its estimation.

1.1   A practical design problem

As an illustration of a practical design problem, consider an agricultural experiment carried out at the Institut National de la Recherche Agronomique (INRA) in France and described in Cliquet, Durier and Kobilinsky (1994). The purpose of the experiment was to improve the production of a bacterium, called bradyrhizobium japonicum, needed for growing soybeans in France. Inoculation is usually performed with sterile peat containing a culture of the bacterium. The peat is mixed with the soybean seeds on sowing. The problem with this technique is that the bacterial survival is severely reduced by dessication in the hours following inoculation. Therefore, a high bacterial density is required during growth and storage. It is thus necessary to use a culture medium that allows for a high bacterial density. The purpose of the experiment was to find such a medium.

The factors studied in the experiment and their levels were determined by preliminary trials and from bibliographic data. Seven factors were selected: carbon (C) source, carbon dose, organic nitrogen (N) source, nitrogen dose, yeast extract dose, ammonium chloride (NH_4Cl) dose and acidity. All doses were measured in gram per liter (g/l) and the acidity was measured using the pH. Two of the factors in the experiment, carbon source and nitrogen source, were qualitative. The factor carbon source possessed four levels:

Table 1.1: Factors and factor levels in the INRA experiment.

Factor	Number of levels	Levels			
Carbon source	4	Mannitol	Glycerol	Gluconate	Glucose
Carbon dose (in g/l)	2		4	6	
Organic nitrogen source	2			Casein hydrolysate	Sodium glutamate
Organic nitrogen dose (in g/l)	4	0.1	0.2	0.3	0.4
Yeast extraction dose (in g/l)	4	1	2	3	4
Ammonium chloride dose (in g/l)	2			0.0	0.1
Acidity (in pH)	2		6	7	

mannitol, glycerol, gluconate and glucose. The factor nitrogen source had only two levels: casein hydrolysate and sodium glutamate. The five remaining factors in the study were quantitative. Two of them, organic nitrogen dose and yeast extract dose, were studied at four levels in order to determine curvature. For example, yeast extract, an important growth factor, might show toxicity at high concentrations, so that the effect of increasing the yeast extract dose might depend on its level. The three other quantitative factors were studied at two levels only. The factor levels considered by the researchers are displayed in Table 1.1.

As a result, $4 \times 4 \times 4 \times 2 \times 2 \times 2 \times 2 = 1024$ factor level combinations or treatments were available. Of course, it would have been too expensive to perform 1024 test runs so that only a small fraction of the treatments could actually be performed. The problem of designing the INRA experiment therefore consists of choosing the best possible treatment for each experimental run. Of course, the choices made should lead to a highly informative experiment. Cliquet et al. (1994) used a fractional factorial design to select the 64 factor level combinations used in the experiment.

From the 64 test runs, it turned out that the factors carbon source, organic nitrogen source, organic nitrogen dose, yeast extract dose and acidity had a significant impact on the amount of bacteria produced. The two other factors, carbon dose and ammonium chloride dose, had no significant impact on the density of the bacteria. Using the information from the experiment, the researchers were able to select two media that allowed more than 10^{10} bacteria per milliliter.

1.2   Analysis of experiments

The main purpose of experimenting is to determine the relationship between the value of one or more dependent or response variables and the settings of one or more independent or explanatory variables. This relationship is quantified by means of a mathematical model and is obtained by the statistical analysis of the experimental data. We will restrict ourselves to the case of a single response variable.

1.2.1   Theory

In the statistical model, the response variable is usually denoted by y, whereas the m explanatory variables are typically represented by $x_1, x_2, \ldots, x_m$. The dependence of the response upon the levels of the explanatory factors is modelled by the response function

$$y = \mathbf{f}'(\mathbf{x})\boldsymbol{\beta}, \qquad\qquad (1.1)$$

where $\mathbf{f}$ is the polynomial expansion of the explanatory variables $\mathbf{x} = [\ x_1\ x_2\ \ldots\ x_m\ ]'$ and $\boldsymbol{\beta}$ is the $p \times 1$ vector containing the parameters of the explanatory variables. Because the experimental observations are subject to random variation, the statistical model adds a random error term ε to the response function. The ith experimental observation can then be written as

$$y_i = \mathbf{f}'(\mathbf{x}_i)\boldsymbol{\beta} + \varepsilon_i. \qquad\qquad (1.2)$$

In this expression, $\mathbf{x}_i$ represents the settings of the explanatory variables in the ith experimental run and is referred to as the design point or treatment corresponding to the ith observation.

Over n observations, this model can be expressed in matrix notation as

$$\mathbf{y} = \mathbf{X}\boldsymbol{\beta} + \boldsymbol{\varepsilon}, \qquad\qquad (1.3)$$

where $\mathbf{y}$ is the $n \times 1$ vector of responses and $\mathbf{X}$ is the $n \times p$ extended design matrix, or simply the design matrix, containing the settings of the explanatory variables in each experimental run. The random error terms are usually assumed to be independent and identically distributed with zero mean and variance σ_ε^2, that is

$$E(\varepsilon_i) = 0, \tag{1.4}$$

$$\text{cov}(\varepsilon_i, \varepsilon_j) = 0, \tag{1.5}$$

$$\text{var}(\varepsilon_i) = \sigma_\varepsilon^2. \tag{1.6}$$

In the sequel of this text, we will refer to the model defined by the equations (1.3) to (1.6) as the uncorrelated model. The assumption (1.5) is realistic in cases where all the runs of the experiment have been randomized and in which the levels of the experimental factors have been reset independently for each run. This kind of experiment is referred to as a completely randomized design. For tests of significance (e.g. t- and F-tests) to be valid, the random errors need to be normally distributed:

$$\boldsymbol{\varepsilon} \sim \mathcal{N}(\mathbf{0}_n, \sigma_\varepsilon^2 \mathbf{I}_n), \tag{1.7}$$

where $\mathbf{0}_n$ is an n-dimensional column vector of zeros and $\mathbf{I}_n$ is the n-dimensional identity matrix.

Unbiased estimates of the unknown model parameters $\boldsymbol{\beta}$ are obtained from the ordinary least squares or OLS estimator

$$\hat{\boldsymbol{\beta}} = (\mathbf{X}'\mathbf{X})^{-1}\mathbf{X}'\mathbf{y}, \tag{1.8}$$

which is equivalent to the maximum likelihood estimator under normal errors. The variance-covariance matrix can be expressed as

$$\text{var}(\hat{\boldsymbol{\beta}}) = \sigma_\varepsilon^2 (\mathbf{X}'\mathbf{X})^{-1}, \tag{1.9}$$

and the information matrix on the unknown fixed parameter $\boldsymbol{\beta}$ is given by

$$\mathbf{M} = \sigma_\varepsilon^{-2} \, \mathbf{X}'\mathbf{X}. \tag{1.10}$$

The information matrix is called regular if its determinant is strictly positive. It is singular if its determinant is zero. A design or the corresponding information matrix is singular if the number of observations is smaller than the number of model parameters p, if the number of distinct design points is smaller than p or if the number of factor levels is too small. A singular design does not allow estimation of $\boldsymbol{\beta}$. A thorough discussion of information matrices can be found in Pukelsheim (1993).

Once the model is estimated, it is used to predict the response for several combinations $\mathbf{x}$ of the experimental factors. The predicted response is given by

$$\hat{y}(\mathbf{x}) = \mathbf{f}'(\mathbf{x})\hat{\boldsymbol{\beta}}, \tag{1.11}$$

and the variance can be written as

$$\text{var}\{\hat{y}(\mathbf{x})\} = \sigma_\varepsilon^2 \, \mathbf{f}'(\mathbf{x})(\mathbf{X}'\mathbf{X})^{-1}\mathbf{f}(\mathbf{x}). \tag{1.12}$$

Table 1.2: Gas turbine experiment described by Myers and Montgomery (1995).

Run	Voltage	Blade Speed (inch/second)	Extension (inch)
1	1.23	5300	0.000
2	3.13	8300	0.000
3	1.22	5300	0.012
4	1.92	8300	0.012
5	2.02	6800	0.000
6	1.51	6800	0.012
7	1.32	5300	0.006
8	2.62	8300	0.006
9	1.65	6800	0.006
10	1.62	6800	0.006
11	1.59	6800	0.006

Since the variance component σ_ε^2 is usually unknown, it has to be estimated. An unbiased estimator is

$$\hat{\sigma}_\varepsilon^2 = \frac{\mathbf{r}'\mathbf{r}}{n-p}, \tag{1.13}$$

where $\mathbf{r}$ is the n-dimensional vector containing the residuals $r_i = y_i - \mathbf{f}'(\mathbf{x}_i)\hat{\boldsymbol{\beta}}$.

From (1.9) and (1.10), it is clear that the properties of the parameter estimates are influenced by the settings of the explanatory variables. The purpose of statistical design of experiments is to determine those settings that generate the best possible estimates.

1.2.2 Illustration

Myers and Montgomery (1995) describe an experiment dealing with gas turbine engines. Voltage output of engines was measured at various combinations of blade speed and voltage measuring sensor extension. The data of the experiment are given in Table 1.2. The purpose of the experiment was to estimate a full quadratic model with voltage output as the dependent variable and blade speed and voltage measuring sensor extension as the explanatory variables.

Coded variables

It is convenient for most applications to describe the experiment in terms of coded variables because this facilitates the comparison of designs from different experiments. Therefore, the quantitative variables are rescaled. It

Table 1.3: Coded form of the gas turbine experiment.

Run	Voltage	Blade Speed	Extension
1	1.23	−1	−1
2	3.13	+1	−1
3	1.22	−1	+1
4	1.92	+1	+1
5	2.02	0	−1
6	1.51	0	+1
7	1.32	−1	0
8	2.62	+1	0
9	1.65	0	0
10	1.62	0	0
11	1.59	0	0

is characteristic for a quantitative or continuous variable u that it varies between a minimum and a maximum value, u_{min} and u_{max}. Typically, the factor levels are rescaled to lie between -1 and +1. The coded values can then be computed by

$$x = \frac{u - u_0}{\Delta}, \qquad (1.14)$$

where u_0 is the midpoint of the interval $[u_{min}, u_{max}]$ and Δ is half the difference between u_{max} and u_{min}. For the interpretation of the experimental results, however, it is desirable to return to the original factor levels. For the gas turbine experiment, the coded levels can be obtained in the following fashion:

$$x_1 = \frac{\text{Blade speed} - 6800}{1500},$$

and

$$x_2 = \frac{\text{Extension} - 0.006}{0.006},$$

where x_1 and x_2 represent the coded levels of the factors blade speed and voltage measuring sensor extension respectively. The coded levels are displayed in Table 1.3. We will use this form to analyze the data. From Table 1.3, it is easy to see that the runs 9, 10 and 11 are carried out at the middle level of the experimental factors. These runs are therefore referred to as center runs.

Analysis

The purpose of the experiment was to estimate a full quadratic model in the two variables. As a result, the polynomial expansion

$$\mathbf{f'(x)} = \begin{bmatrix} 1 & x_1 & x_2 & x_1x_2 & x_1^2 & x_2^2 \end{bmatrix},$$

and

$$\boldsymbol{\beta'} = \begin{bmatrix} \beta_0 & \beta_1 & \beta_2 & \beta_{12} & \beta_{11} & \beta_{22} \end{bmatrix},$$

so that $p = 6$ and the statistical model can be written as

$$y = \beta_0 + \beta_1 x_1 + \beta_2 x_2 + \beta_{12} x_1 x_2 + \beta_{11} x_1^2 + \beta_{22} x_2^2 + \varepsilon.$$

The (extended) design matrix for the entire experiment is given by

$$\mathbf{X} = \begin{bmatrix}
1 & -1 & -1 & +1 & +1 & +1 \\
1 & +1 & -1 & -1 & +1 & +1 \\
1 & -1 & +1 & -1 & +1 & +1 \\
1 & +1 & +1 & +1 & +1 & +1 \\
1 & 0 & -1 & 0 & 0 & +1 \\
1 & 0 & +1 & 0 & 0 & +1 \\
1 & -1 & 0 & 0 & +1 & 0 \\
1 & +1 & 0 & 0 & +1 & 0 \\
1 & 0 & 0 & 0 & 0 & 0 \\
1 & 0 & 0 & 0 & 0 & 0 \\
1 & 0 & 0 & 0 & 0 & 0
\end{bmatrix}.$$

The first column of $\mathbf{X}$ is a column of ones and corresponds to the intercept. The second and third column contain the settings of x_1 and x_2 at each run of the experiment. The fourth column corresponds to the interaction of both explanatory variables and is obtained by multiplying the levels of x_1 and x_2. Finally, the fifth and sixth column correspond to the quadratic effects of x_1 and x_2 respectively and are obtained by taking the square of their levels. We have that

$$\mathbf{X'X} = \begin{bmatrix}
11 & 0 & 0 & 0 & 6 & 6 \\
0 & 6 & 0 & 0 & 0 & 0 \\
0 & 0 & 6 & 0 & 0 & 0 \\
0 & 0 & 0 & 4 & 0 & 0 \\
6 & 0 & 0 & 0 & 6 & 4 \\
6 & 0 & 0 & 0 & 4 & 6
\end{bmatrix},$$

and hence,

$$(\mathbf{X'X})^{-1} = \begin{bmatrix}
0.2632 & 0 & 0 & 0 & -0.1579 & -0.1579 \\
0 & 0.1667 & 0 & 0 & 0 & 0 \\
0 & 0 & 0.1667 & 0 & 0 & 0 \\
0 & 0 & 0 & 0.25 & 0 & 0 \\
-0.1579 & 0 & 0 & 0 & 0.3947 & -0.1053 \\
-0.1579 & 0 & 0 & 0 & -0.1053 & 0.3947
\end{bmatrix}.$$

Since

$$\mathbf{y}' = \begin{bmatrix} 1.23 & 3.13 & 1.22 & 1.92 & \cdots & 2.62 & 1.65 & 1.62 & 1.59 \end{bmatrix},$$

we also have that

$$\mathbf{X}'\mathbf{y} = \begin{bmatrix} 19.83 & 3.9 & -1.73 & -1.2 & 11.44 & 11.03 \end{bmatrix}',$$

and, as a result, the ordinary least squares estimates for β are given by

$$\hat{\beta} = \begin{bmatrix} \hat{\beta}_0 \\ \hat{\beta}_1 \\ \hat{\beta}_2 \\ \hat{\beta}_{12} \\ \hat{\beta}_{11} \\ \hat{\beta}_{22} \end{bmatrix} = \begin{bmatrix} 1.6705 \\ 0.6500 \\ -0.2883 \\ -0.3000 \\ 0.2237 \\ 0.0187 \end{bmatrix}.$$

The estimated model then becomes

$$y = 1.6705 + 0.6500x_1 - 0.2883x_2 - 0.3000x_1x_2 + 0.2237x_1^2 + 0.0187x_2^2.$$

For inference purposes, an estimate for the variance σ_ε^2 is necessary. We have that the vector of residuals equals

$$\mathbf{r}' = \begin{bmatrix} -0.0212 & -0.0212 & -0.0546 & \cdots & -0.0205 & -0.0505 & -0.0805 \end{bmatrix},$$

and the sum of squared residuals is 0.0415, so that

$$\hat{\sigma}_\varepsilon^2 = \frac{0.0415}{11 - 6} = 0.0083.$$

1.3   Design of experiments

The earliest statistical investigations about the planning of experiments originated almost entirely from agriculture, as can be seen in Fisher (1935). Emphasis was laid on detecting the influences of discrete or qualitative factors on the output and the construction of the experimental designs extensively made use of combinatorial principles. Well-known examples of such designs are Latin squares and balanced incomplete block designs. A whole branch of theory for categorical designs developed along these lines. An excellent overview is given by Cox (1958).

For experimental design problems with continuous factors, a series of standard response surface designs have been proposed. Probably the most famous category of standard designs is the two-level factorial design, which has excellent properties for the estimation of first order models. Other first order standard response surface designs are the two-level fractional factorial designs, the simplex designs and the designs developed by Plackett and Burman (1946). For second order models, Box and Wilson (1951) introduced the popular central composite designs. Other design options for second order models are the three-level factorial designs and the designs described by Box and Behnken (1960). A common aim of these designs for regularly shaped design regions is to comply with Box and Draper's (1971) list of requirements for proper experimentation:

1. Generate a satisfactory distribution of information throughout the region of interest.

2. Ensure that the fitted values are as close as possible to the true values of the response.

3. Allow detection of lack-of-fit.

4. Allow estimation of transformations of both the response and the quantitative experimental factors.

5. Allow blocked experiments.

6. Allow designs to be built up sequentially.

7. Provide an estimate of error from replication.

8. Be insensitive to wild observations and to violation of normality assumptions.

9. Require a minimum of experimental runs.

10. Provide simple data patterns and allow visual appreciation.

11. Ensure simplicity of calculation.

12. Behave well when errors occur in the settings of the explanatory variables.

13. Avoid large numbers of different factor levels.

14. Provide a check for the assumption of homogeneous variance.

It is evident that experimental designs satisfying all these conditions are rare and that some of the requirements have become less important thanks to the widespread use of the computer. Moreover, some of the requirements may be more important than others, depending on the experimental situation. For instance, if experiments are to be conducted by unskilled workers, large numbers of factor levels should be avoided as much as possible.

A third branch of experimental design theory is concerned with the optimal design of experiments. The aim of the theory of optimal designs is to plan the experiment so that it provides the experimenter with a maximum of information on the model under investigation. An important advantage of this approach is that the resulting design matches the desires of the experimenter. Unlike standard response surface designs, optimal designs can be constructed for any number of observations. In addition, this approach can take into account any possible restriction on the settings of the experimental variables. It is also able to cope with experimental situations with any combination of quantitative, qualitative and mixture variables and it provides a framework for the computation of efficient designs in cases where existing designs provide no alternative, for instance for those combinations of the number of treatments, the number of blocks and the block size where no appropriate categorical design is found in the literature. In other words, optimal design theory allows the researcher to design an experiment for a given situation. When standard response surface designs and categorical designs are used, the experimental situation is often adjusted to the design. The optimal design theory was put forward by Kiefer (1959). Excellent overviews of the work in this area are given by Fedorov (1972), Silvey (1980) and Atkinson and Donev (1992).

Because of these three different approaches to the design of experiments, the literature on the topic is completely scattered and the terminology used is often specific for one particular field. Firstly, we provide the reader with a brief introduction to optimal design. Next, we describe the most popular standard response surface and categorical designs and indicate which of these designs is optimal.

1.4   Optimal designs

In this section, we focus on the optimal design of experiments and explain the difference between continuous designs and discrete designs. The former are only of theoretical importance, while the latter are meant to be applied in practice. We also define the most important design criteria and describe the basic theorem of optimal design theory, the General Equivalence Theorem. Finally, we provide the reader with some examples of optimal designs.

1.4.1   Discrete versus continuous designs

The purpose of optimal design theory is to determine the values of the explanatory variables x each time an experimental observation is made. In the literature, the mathematical problem of finding the optimal design is tackled in two ways. The first approach ignores the fact that the number of observations at a certain design point must be an integer. The resulting designs are called continuous, approximate or asymptotic. Continuous designs are represented by a measure ξ on the design region χ. If the design has observations at h distinct design points $\mathbf{x}_i \in \chi$, it is denoted by

$$\xi = \begin{Bmatrix} \mathbf{x}_1 & \mathbf{x}_2 & \cdots & \mathbf{x}_h \\ w_1 & w_2 & \cdots & w_h \end{Bmatrix}, \tag{1.15}$$

where the first row gives the levels of the experimental factor in each design point and the second row gives the weights associated with each design point. Since ξ is a measure, $\int_\chi \xi(d\mathbf{x}) = 1$ and $0 \leq w_i \leq 1$ for all i. For a continuous design, the information matrix is defined as

$$\mathbf{M}(\xi) = \sum_{i=1}^{h} w_i \mathbf{f}(\mathbf{x}_i)\mathbf{f}'(\mathbf{x}_i), \tag{1.16}$$

and the standardized prediction variance is defined as

$$d(\mathbf{x}, \xi) = \mathbf{f}'(\mathbf{x})\{\mathbf{M}(\xi)\}^{-1}\mathbf{f}(\mathbf{x}). \tag{1.17}$$

The second approach to optimal design takes into account the integrality constraint. The resulting designs are called discrete or exact. A discrete design with n observations can be denoted as

$$\xi_n = \begin{Bmatrix} \mathbf{x}_1 & \mathbf{x}_2 & \cdots & \mathbf{x}_h \\ n_1 & n_2 & \cdots & n_h \end{Bmatrix}, \tag{1.18}$$

where n_i is the number of observations at design point $\mathbf{x}_i$ and $\sum_{i=1}^{h} n_i = n$. The information matrix of a discrete design is given by

$$\mathbf{M} = \sigma_\varepsilon^{-2} \sum_{i=1}^{h} n_i \mathbf{f}(\mathbf{x}_i)\mathbf{f}'(\mathbf{x}_i), \tag{1.19}$$

which is equal to (1.10). The standardized prediction variance for a discrete design is defined as

$$\begin{aligned} d(\mathbf{x}, \xi_n) &= n\sigma_\varepsilon^{-2}\mathbf{f}'(\mathbf{x})\mathbf{M}^{-1}\mathbf{f}(\mathbf{x}), \\ &= n\sigma_\varepsilon^{-2}\mathrm{var}\{\hat{y}(\mathbf{x})\}. \end{aligned} \tag{1.20}$$

In practice all designs are of course discrete. For moderate and large n, efficient —though not necessarily optimal— discrete designs can frequently be found by multiplying the weights w_i of the optimal continuous design by n and rounding this product to the nearest integer. In some specific instances, this approach of rounding the optimal continuous design produces the optimal discrete design.

1.4.2 Optimality criteria

Numerous optimality criteria can be found in the literature. Many of them belong to the class of alphabetic optimality criteria because they are named after a letter. The most important design criterion in applications is the $\mathcal{D}$-optimality criterion. Its advantages relative to the other criteria are explained later in this chapter. The optimality criteria described here are all functions Ψ of the information matrix $\mathbf{M}$. Each of them allows the experimenter to order alternative design options. Unfortunately, the ordering of designs depends on the criterion used. The ordering of two designs ξ_I and ξ_{II} is however independent of the design criterion if $\mathbf{M}(\xi_I) - \mathbf{M}(\xi_{II})$ is positive definite. In that case, ξ_I is a better design than ξ_{II} for any generalized optimality criterion. The class of generalized optimality criteria includes the $\mathcal{A}$-, $\mathcal{D}$- and $\mathcal{E}$-optimality criterion described below. It was introduced by Kiefer (1975) and studied in detail by Cheng (1978). Of course, the optimal design ξ^* with respect to a given criterion is the design that optimizes the criterion value over the space Ξ of all feasible designs. Mathematically,

$$\Psi\{\mathbf{M}(\xi^*)\} = \underset{\xi \in \Xi}{\mathrm{opt}} \ \Psi\{\mathbf{M}(\xi)\}. \tag{1.21}$$

We now define some of the most popular design criteria in the literature. For notational simplicity, we have omitted the constant σ_ε^2, which is equivalent to assuming that $\sigma_\varepsilon^2 = 1$.

$\mathcal{D}$-optimality

The $\mathcal{D}$-optimal design minimizes the generalized variance of the parameter estimators. This is done by minimizing the determinant of the variance-covariance matrix of the parameter estimators or, equivalently, by maximizing the determinant of the information matrix. The $\mathcal{D}$-optimal design is thus obtained by maximizing the determinant

$$|\mathbf{X}'\mathbf{X}|. \tag{1.22}$$

Denoting the p eigenvalues of the information matrix by $\lambda_1, \lambda_2, \ldots, \lambda_p$, the $\mathcal{D}$-criterion value (1.22) can also be written as

$$\prod_{i=1}^{p} \lambda_i. \tag{1.23}$$

The motivation for the $\mathcal{D}$-optimality criterion lies in the fact that the $\mathcal{D}$-optimal design minimizes the volume of the confidence ellipsoid about the unknown parameters $\boldsymbol{\beta}$.

$\mathcal{D}_s$-optimality

When the interest is in estimating a subset of the model parameters as precisely as possible, this should be taken into account when designing the experiment. Let us rewrite the regression model (1.3) as follows:

$$\mathbf{y} = \mathbf{X}_1\boldsymbol{\beta}_1 + \mathbf{X}_2\boldsymbol{\beta}_2 + \boldsymbol{\varepsilon}, \tag{1.24}$$

where $\boldsymbol{\beta}_1$ contains the s parameters of interest, $\boldsymbol{\beta}_2$ contains the remaining $p - s$ model parameters and $\mathbf{X}_1$ and $\mathbf{X}_2$ represent the corresponding parts of the design matrix $\mathbf{X}$. The information matrix can then be written as

$$\mathbf{X'X} = \begin{bmatrix} \mathbf{X}_1'\mathbf{X}_1 & \mathbf{X}_1'\mathbf{X}_2 \\ \mathbf{X}_2'\mathbf{X}_1 & \mathbf{X}_2'\mathbf{X}_2 \end{bmatrix}. \tag{1.25}$$

The variance-covariance matrix of the least squares estimator $\hat{\boldsymbol{\beta}}_1$ of $\boldsymbol{\beta}_1$ is given by the upper left submatrix of $(\mathbf{X'X})^{-1}$. Applying Theorem 8.5.11 of Harville (1997), this submatrix can be written as

$$\{\mathbf{X}_1'\mathbf{X}_1 - \mathbf{X}_1'\mathbf{X}_2(\mathbf{X}_2'\mathbf{X}_2)^{-1}\mathbf{X}_2'\mathbf{X}_1\}^{-1}. \tag{1.26}$$

The $\mathcal{D}_s$-optimal design minimizes the determinant of this variance-covariance matrix, or equivalently, it maximizes its inverse

$$|\mathbf{X}_1'\mathbf{X}_1 - \mathbf{X}_1'\mathbf{X}_2(\mathbf{X}_2'\mathbf{X}_2)^{-1}\mathbf{X}_2'\mathbf{X}_1|.$$

Using Harville's (1997) Theorem 13.3.8, we have that

$$|\mathbf{X}_1'\mathbf{X}_1 - \mathbf{X}_1'\mathbf{X}_2(\mathbf{X}_2'\mathbf{X}_2)^{-1}\mathbf{X}_2'\mathbf{X}_1| = \frac{|\mathbf{X'X}|}{|\mathbf{X}_2'\mathbf{X}_2|}. \tag{1.27}$$

The $\mathcal{D}_s$-optimal design for estimating $\tilde{\boldsymbol{\beta}}$, where $\tilde{\boldsymbol{\beta}}$ contains all elements of $\boldsymbol{\beta}$ except the intercept, turns out to be equivalent to the $\mathcal{D}$-optimal design. In order to see this, substitute $\mathbf{X}_1 = \tilde{\mathbf{X}}$ and $\mathbf{X}_2 = \mathbf{1}_n$ in the previous equation to obtain

$$|\tilde{\mathbf{X}}'\tilde{\mathbf{X}} - \tilde{\mathbf{X}}'\mathbf{1}_n(\mathbf{1}_n'\mathbf{1}_n)^{-1}\mathbf{1}_n'\tilde{\mathbf{X}}| = \frac{|\mathbf{X'X}|}{|\mathbf{1}_n'\mathbf{1}_n|},$$

and, since $\mathbf{1}_n'\mathbf{1}_n = n$,

$$|\tilde{\mathbf{X}}'\tilde{\mathbf{X}} - n^{-1}\tilde{\mathbf{X}}'\mathbf{1}_n\mathbf{1}_n'\tilde{\mathbf{X}}| = \frac{|\mathbf{X'X}|}{n}. \tag{1.28}$$

As a result, the $\mathcal{D}$-criterion value is a constant multiple of the $\mathcal{D}_s$-criterion value. Therefore, both criteria are equivalent when the interest is in estimating all effects apart from the intercept. Note that the variance-covariance matrix of $\hat{\tilde{\boldsymbol{\beta}}}$ is given by

$$\text{var}(\hat{\tilde{\boldsymbol{\beta}}}) = \sigma_\varepsilon^2(\tilde{\mathbf{X}}'\tilde{\mathbf{X}} - n^{-1}\tilde{\mathbf{X}}'\mathbf{1}_n\mathbf{1}_n'\tilde{\mathbf{X}})^{-1}. \tag{1.29}$$

A similar equivalence between $\mathcal{D}$- and $\mathcal{D}_s$-optimal designs is derived for the fixed block effects model in Section 2.3.6.

$\mathcal{A}$-optimality

The $\mathcal{A}$-optimal design minimizes the sum or the average of the variances of the parameter estimators. The $\mathcal{A}$-optimality criterion value can be written as

$$\sum_{i=1}^{p} \text{var}(\hat{\beta}_i), \tag{1.30}$$

$$\text{tr}(\mathbf{X}'\mathbf{X})^{-1}, \tag{1.31}$$

or, equivalently, as

$$\sum_{i=1}^{p} \frac{1}{\lambda_i}. \tag{1.32}$$

An advantage of the $\mathcal{A}$-optimality criterion with respect to other design criteria is that the coefficients can be weighted. A drawback of the $\mathcal{A}$-optimality criterion is that the optimal design depends on the scale used to measure the experimental factors and on the parameterization used (see also Section 1.7).

$\mathcal{E}$-optimality

The $\mathcal{E}$-optimal design minimizes the variance of the least well-estimated contrast $\mathbf{a}'\boldsymbol{\beta}$ with $\mathbf{a}'\mathbf{a} = 1$. The $\mathcal{E}$-criterion value is given by

$$\max_{i=1,\dots,p} \frac{1}{\lambda_i}. \tag{1.33}$$

$\mathcal{G}$-optimality

The $\mathcal{G}$-optimal design minimizes the maximum of the prediction variance over the design region χ. The $\mathcal{G}$-criterion value is given by

$$\max \text{var} \{\hat{y}(\mathbf{x})\} = \max \mathbf{f}'(\mathbf{x})(\mathbf{X}'\mathbf{X})^{-1}\mathbf{f}(\mathbf{x}) \tag{1.34}$$

For continuous designs, this design criterion is equivalent to the $\mathcal{D}$-optimality criterion. This is a consequence of the General Equivalence Theorem (see Section 1.4.3).

$\mathcal{V}$-optimality

The $\mathcal{V}$-optimal design minimizes the average prediction variance over the design region χ. This criterion is sometimes referred to as $\mathcal{Q}$-optimality, as $\mathcal{I}$-optimality or as $\mathcal{IV}$-optimality as well. The $\mathcal{V}$-criterion value is given by

$$\mathcal{V} = \int_{\chi} \mathbf{f}'(\mathbf{x})(\mathbf{X}'\mathbf{X})^{-1}\mathbf{f}(\mathbf{x})d\mathbf{x}. \tag{1.35}$$

1.4.3   The General Equivalence Theorem

In the theory of optimal continuous designs, a certain function of the information matrix $\mathbf{M}$, say $\boldsymbol{\Psi}(\mathbf{M})$, is minimized. The General Equivalence Theorem states that the following three conditions are equivalent:

1. the optimal design ξ^* minimizes $\boldsymbol{\Psi}$;

2. the minimum of the derivative of Ψ evaluated in $\xi^* \geq 0$;

3. the derivative of Ψ evaluated in ξ^* achieves its minimum at the design points.

The theorem assumes that the design region χ is compact and that the function Ψ is convex and differentiable. As a result, any minimum found will certainly be global. One design criterion that satisfies these conditions is the $\mathcal{D}$-optimality criterion, in which $\Psi(\mathbf{M}) = -\ln|\mathbf{M}|$ and the derivative Ψ' is given by

$$\Psi'(\mathbf{x}, \xi) = p - d(\mathbf{x}, \xi), \tag{1.36}$$

where $d(\mathbf{x}, \xi)$ is the standardized variance defined in (1.17). A direct consequence of the theorem is that $\mathcal{D}$-optimal continuous designs are also $\mathcal{G}$-optimal, i.e. they minimize the maximum of the prediction variance over the design region. Unfortunately, the General Equivalence Theorem is in general not valid for discrete designs. It can be used, however, to prove the $\mathcal{D}$-optimality of the two-level factorial designs and the Plackett-Burman designs (see Section 1.5.1).

1.4.4   Some illustrations

We will now present some examples of continuous and discrete optimal designs.

Continuous designs

The $\mathcal{D}$- and $\mathcal{G}$-optimal continuous designs for polynomial regression on one variable $x \in \chi = [-1, 1]$ depend upon the derivative of the Legendre polynomial $P_d(x)$. Guest (1958) and Hoel (1958) extended the results of Smith (1918) by showing that the $d+1$ points of support of the $\mathcal{D}$-optimal design for the dth order polynomial in one explanatory variable x

$$y = \beta_0 + \sum_{i=1}^{d} \beta_i x^i + \varepsilon \tag{1.37}$$

are at ± 1 and the roots of the equation

$$P_d'(x) = 0. \tag{1.38}$$

All design points have equal weight $(d+1)^{-1}$. The set of Legendre polynomials can be found in Abramowitz and Stegun (1970). The optimal x values up to the sixth order polynomial are displayed in Table 1.4. Because of the equivalency between the $\mathcal{D}$- and the $\mathcal{G}$-optimality criterion for continuous designs, the $\mathcal{D}$-optimal designs are $\mathcal{G}$-optimal as well.

Table 1.4: $\mathcal{D}$-optimal design points for the dth order polynomial regression in one variable.

d	x_1	x_2	x_3	x_4	x_5	x_6	x_7
1	-1						1
2	-1			0			1
3	-1		$-a_3$		a_3		1
4	-1		$-a_4$	0	a_4		1
5	-1	$-a_5$	$-b_5$		b_5	a_5	1
6	-1	$-a_6$	$-b_6$	0	b_6	a_6	1

$$a_3 = \sqrt{1/5}$$
$$a_4 = \sqrt{3/7}$$
$$a_5 = \sqrt{(7 + 2\sqrt{7})/21}$$
$$b_5 = \sqrt{(7 - 2\sqrt{7})/21}$$
$$a_6 = \sqrt{(15 + 2\sqrt{15})/33}$$
$$b_6 = \sqrt{(15 - 2\sqrt{15})/33}$$

The $\mathcal{A}$-optimal design for quadratic regression on one variable is given by

$$\xi = \left\{ \begin{matrix} -1 & 0 & 1 \\ 1/4 & 1/2 & 1/4 \end{matrix} \right\}. \tag{1.39}$$

For second order polynomials in m factors

$$y = \beta_0 + \sum_{i=1}^{m} \beta_i x_i + \sum_{i=1}^{m-1} \sum_{j=i+1}^{m} \beta_{ij} x_i x_j + \sum_{i=1}^{m} \beta_{ii} x_i^2 + \varepsilon, \tag{1.40}$$

continuous $\mathcal{D}$-optimal designs are given by Farrell, Kiefer and Walbran (1967). For a spherical design region, the optimal designs have weight $2/\{(m+1)(m+2)\}$ at the center point and the rest of the weight is uniformly spread over the sphere of radius $\sqrt{m}$. It should be noted at this point that the central composite designs do not provide the optimal design points for the continuous $\mathcal{D}$-optimal designs. For a cubic design region, the optimal continuous designs are supported on three subsets of the points of the 3^m factorial. A $\mathcal{D}$-optimal design is supported on the center point, the 2^m corner points and the midpoint of the edges. For $m = 2$, the center point has weight 0.096, while the corner points have weight 0.14575 each and the midpoints of the edges have weight 0.08025 each. Further details and references can be found in Atkinson and Donev (1992).

Finally, let us consider $\mathcal{D}$-optimal designs for a mixture experiment. In mixture experiments, the response only depends on the proportions of the components but not on the total amount. Of course, each proportion lies between zero and one and all proportions sum to one. Usually, a canonical polynomial of Scheffé (1958) is used to model this type of experiment. For

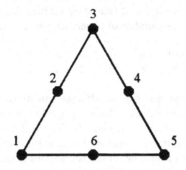

Figure 1.1: Second order lattice design for $m = 3$.

example, the second-order model for an m component mixture is given by

$$y = \sum_{i=1}^{m} \beta_i x_i + \sum_{i=1}^{m-1} \sum_{j=i+1}^{m} \beta_{ij} x_i x_j + \varepsilon, \qquad (1.41)$$

where

$$\sum_{i=1}^{m} x_i = 1 \qquad (x_i \geq 0).$$

Kiefer (1961) established the $\mathcal{D}$-optimality of the first and second order simplex lattice designs for the first and second order Scheffé polynomials respectively. In the $\mathcal{D}$-optimal continuous design, each design point has equal weight and the number of design points equals the number of model parameters. The simplex lattice design of the dth order contains all possible combinations of experimental runs in which each component takes values 0, $1/d$, $2/d, \ldots$, 1. As an illustration, Figure 1.1 shows the design points of the second order lattice design for a three component mixture. For more information on mixture experiments, we refer the interested reader to Cornell (1990).

Discrete designs

The optimality of the continuous designs described above has been proven. In contrast, computer searches are usually the only means to demonstrate the optimality of a discrete design. However, if the weights of the optimal continuous design are rational numbers, then it can be used to construct an optimal discrete design. For example, a 12-point $\mathcal{D}$-optimal design for estimating a 5th order polynomial in one explanatory variable can be constructed from the $\mathcal{D}$-optimal continuous design. The optimal continuous design has equal weight on six distinct design points (see Table 1.4):

$$\left\{ \begin{matrix} -1 & -a_5 & -b_5 & b_5 & a_5 & 1 \\ 1/6 & 1/6 & 1/6 & 1/6 & 1/6 & 1/6 \end{matrix} \right\}. \qquad (1.42)$$

The optimal 12-point design is obtained by multiplying the weights of the continuous design by the number of observations, $n = 12$:

$$\left\{ \begin{matrix} -1 & -a_5 & -b_5 & b_5 & a_5 & 1 \\ 2 & 2 & 2 & 2 & 2 & 2 \end{matrix} \right\}. \tag{1.43}$$

In a similar fashion, optimal n-point designs can be constructed from the corresponding optimal continuous designs provided nw_i is an integer for all design points of the continuous design. If the weights of the optimal designs are not rational numbers, this is of course impossible unless $n \rightarrow \infty$. Kiefer (1971) suggested rounding off the continuous designs and noted that, for large n, the discrete designs obtained in this way are nearly $\mathcal{D}$-optimal.

Usually, the discrete optimal designs are found by searching over the design region χ. In simple problems, analytical solutions are sometimes possible. Mostly, however, numerical methods have to be used. Box and Draper (1971) and Mitchell (1974b) show that, for a cubic design region, the discrete $\mathcal{D}$-optimal designs for a main effects model have observations in the corner points only. This result greatly simplifies the search for an optimal design for this model. For second order models in two and three explanatory variables with χ a square and a cube respectively, Box and Draper (1971) used a quasi-Newton method to compute discrete $\mathcal{D}$-optimal designs. In the case where there are two experimental factors, the discrete optimal designs for $n = 6, \ldots, 9$ are as follows:

$n = 6$: $(-1, -1), (1, -1), (-1, 1), (-\alpha, -\alpha), (1, 3\alpha), (3\alpha, 1)$,
            where $\alpha = (4 - \sqrt{13})/3$,
$n = 7$: $(\pm 1, \pm 1), (-0.092, 0.092), (1, -0.067), (0.067, -1)$,
$n = 8$: $(\pm 1, \pm 1), (1, 0), (0.082, 1), (0.082, -1), (-0.215, 0)$,
$n = 9$: $(\pm 1, \pm 1), (\pm 1, 0), (0, \pm 1)$.

Each of these designs remains optimal if it is rotated through $\pi/2$, π or $3\pi/2$. Their geometric representation is given in Figure 1.2. From the figure, it can be seen that, except for the smallest design, the optimal design points lie close to fractions of the 3^2 factorial design (see Section 1.5.2). Therefore, for second order models, the search over a continuous design region is often replaced by a search over the points of the 3^m factorial. For example, the $\mathcal{D}$-optimal designs for the second order model in two variables, obtained by using the points of the 3^2 factorial as the set of candidates, are displayed in Figure 1.3. More generally, for practical design problems, a list of candidate points is used instead of searching the entire design region. In that case, the design problem is that of selecting n design points out of a list of candidates. Since replication of design points is allowed, the selection is with replacement. If there are N candidate points, the number of possible n-point designs therefore amounts to N^n.

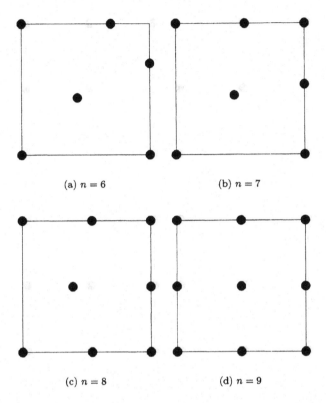

Figure 1.2: Discrete $\mathcal{D}$-optimal designs for the second order model in two factors for a square design region.

1.4.5   Advantages and disadvantages of optimal design theory

The main advantage of optimal design theory is that it allows the researcher to design a tailor-made experiment in an objective way. Optimal designs can be computed for any number of observations, for any number of explanatory variables, for any statistical model, for any combination of qualitative and quantitative variables and for any experimental region. When the experimental units are heterogeneous and the experiment has to be blocked, optimal designs can be computed with any number of blocks and with any block size. Therefore, the theory of optimal design is flexible enough to cope with the wide variety of experimental situations that occur in any field of scientific endeavor. The standard response surface designs introduced in Section 1.5 can be used for experiments with quantitative factors and with one specific number of observations only. In addition, they cannot be used when the experimental region is restricted. Similarly, the categor-

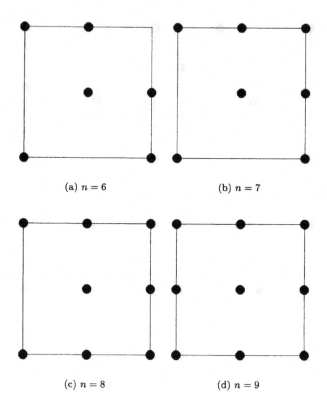

(a) $n = 6$ (b) $n = 7$

(c) $n = 8$ (d) $n = 9$

Figure 1.3: $\mathcal{D}$-optimal three-level designs for the second order model in two factors for a square design region.

ical designs described in Section 1.6 can only be used for specific numbers of observations, blocks and treatments. Of course, the standard response surface designs as well as the categorical designs can be adapted in order to meet with the experimenter's wishes. However, this adaptation is seldom guided by objective criteria and therefore does not always lead to a good design.

The optimal design approach also allows the experimenter to augment an experiment, that is to design an experiment incorporating existing data. This is very important because experiments are often conducted in stages. In the first stage, a screening experiment is usually carried out to estimate a simple first order model. The primary goal of this stage is to identify the important factors. In the next stages, additional experimental runs are performed to estimate a more complex model. The best possible choice of the runs in these stages of course depends on the runs performed in the previous stages.

One drawback of the optimal design theory is that the statistical model has to be specified in advance. In other words, the power expansion f is required as an input to any design construction algorithm. In most cases, a design that is optimal for one model will not be optimal for another. In addition, a design that is optimal for one model does sometimes not allow the estimation of another. For example, optimal designs for first order regression models possess only two factor levels. With these designs, no quadratic effects can be estimated because this requires at least three different factor levels. Of course, a similar drawback is encountered when choosing a standard response surface design. For example, no quadratic effects can be estimated when a first order standard response surface design is used. Similarly, some of the higher order interactions cannot be estimated when a fractional factorial design is used.

Another drawback of optimal designs is that they sometimes do not satisfy Box and Draper's (1971) list of requirements for proper experimentation, given in Section 1.3. More specifically, optimal designs can be saturated, i.e. the number of design points is equal to the number of model parameters p, so that lack-of-fit cannot be detected.

Another unfortunate aspect of the optimal design theory with continuous factors is that the optimal design depends on the design criterion. Blinded by the wide variety of design criteria available, many experimenters fall back on standard designs even though these designs are not always fit for the design problem at hand. However, $\mathcal{D}$-optimal designs often perform well with respect to other criteria (see Section 1.7), so that the $\mathcal{D}$-optimality criterion could help the experimenter out.

Finally, discrete optimal designs are not always symmetric, especially when the number of observations is small. This problem is due to the discreteness of the weights in practical design problems. As a matter of fact, continuous optimal designs usually display symmetry.

1.4.6   Design efficiency

In the literature, the efficiency of a discrete design is usually obtained by comparing the design to the corresponding optimal continuous design for the design problem at hand. Denoting by $\mathbf{M}(\xi^*)$ the information matrix corresponding to the $\mathcal{D}$-optimal continuous design ξ^*, the $\mathcal{D}$-efficiency of an n-point discrete design $\mathbf{X}$ is defined as

$$\mathcal{D}_{\text{eff}} = \left\{ \frac{\left| \frac{1}{n} \mathbf{X}'\mathbf{X} \right|}{|\mathbf{M}(\xi^*)|} \right\}^{1/p}, \tag{1.44}$$

where $|\mathbf{X}'\mathbf{X}/n|$ represents the amount of information per observation provided by $\mathbf{X}$. This efficiency measure is used in the literature to compare designs of different sizes. Similarly, Letsinger, Myers and Lentner (1996) multiply the $\mathcal{V}$- or $\mathcal{Q}$-criterion value by n to compare competing designs of different sizes. Relying on the information per observation is, however, not the best way to compare designs for practical purposes.

In order to demonstrate this, consider two different mixture experiments for the estimation of a second order mixture model. The first experiment is the 6-point lattice design in Figure 1.1. The second experiment is the 7-point design consisting of the 6-point lattice design and one replication of the first point of the lattice. The $\mathcal{D}$-efficiency of the first design is 100% because it can easily be transformed into the $\mathcal{D}$-optimal continuous design by dividing the weights of the design points by six. The $\mathcal{D}$-efficiency of the 7-point design falls to 96%, even though the determinant of its information matrix doubles. The higher $\mathcal{D}$-efficiency of the first design suggests that the 6-point design should be preferred. However, the determinant of the second design is twice as large as that of the first design. As a result, the best estimates can be obtained by using the 7-point design although it has the lowest $\mathcal{D}$-efficiency.

The purpose of this example was to demonstrate that a per observation information value, e.g. the $\mathcal{D}$-efficiency defined in (1.44), should not be used to compare designs. Instead, the comparison should be based on the total information matrix $\mathbf{X}'\mathbf{X}$. In this book, we compare only designs of equal size. This is because the number of experimental observations is usually limited by time or cost constraints. Since additional observations provide extra information, we believe that a maximum number of runs will be carried out given the constraints. This view is shared by Trinca and Gilmour (1999). If designs with different sizes are to be compared, the comparison should be based on the determinants of the total information matrices and not on the per observation information matrices.

1.5   Standard response surface designs

In this section, we give a short overview of the most popular standard response surface designs. They are also referred to as regression designs because they are often analyzed by using regression techniques. Firstly, the first order designs are presented. Secondly, the most popular second order

Table 1.5: 2^3 factorial design.

	x_1	x_2	x_3
1	-1	-1	-1
2	+1	-1	-1
3	-1	+1	-1
4	+1	+1	-1
5	-1	-1	+1
6	+1	-1	+1
7	-1	+1	+1
8	+1	+1	+1

designs are introduced. The geometric and the statistical properties of the designs are described and graphically displayed.

1.5.1   First order designs

First order designs are especially useful for screening experiments. The most popular class of standard response surface designs is undoubtedly the class of 2^m factorial designs, where m is the number of experimental factors. An economical alternative to the full factorial is the fractional factorial design. Plackett and Burman (1946) introduced a class of designs for factors with two levels for values of n which are multiples of four.

2^m factorial designs

A 2^m factorial design consists of all $n = 2^m$ combinations of points at which the factors take the coded values of -1 and $+1$. The most complicated model that can be estimated using this design is a model which contains first order terms in all factors, all two factor interactions and all higher order interactions up to order m. For example, we obtain

$$y = \beta_0 + \sum_{i=1}^{3} \beta_i x_i + \sum_{i=1}^{2} \sum_{j=i+1}^{3} \beta_{ij} x_i x_j + \beta_{123} x_1 x_2 x_3 + \varepsilon \qquad (1.45)$$

for $m = 3$. Note that this model possesses as many parameters as there are observations in the experiment, i.e. $p = n$. Therefore, the design is called saturated. As an illustration, the 2^3 factorial design is given in Table 1.5. The standard order notation has been used. As is shown in Figure 1.4, the design points are the corner points of the experimental region.

The two-level factorial designs are simple to construct and easy to analyze. An important feature is that the design is orthogonal, so that the information matrix $\mathbf{X}'\mathbf{X}$ is diagonal with diagonal elements $n = 2^m$. Orthogonal designs have the advantage that parameter estimates are statistically in-

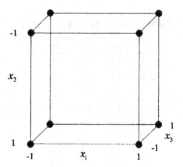

Figure 1.4: 2^3 factorial design.

dependent from each other. In addition, the 2^m factorial is $\mathcal{D}$-optimal provided the design region $\chi = [-1,1]^m$. This can be shown by using the General Equivalence Theorem. For that purpose, we translate the discrete design into a continuous one by giving each design point weight 2^{-m}. Because the design is orthogonal, the information matrix will be the n-dimensional identity matrix. As a result, the standardized prediction variance (1.17) becomes

$$d(\mathbf{x}, \xi) = \mathbf{f}'(\mathbf{x})\mathbf{f}(\mathbf{x})$$
$$= 1 + \sum_{i=1}^{m} x_i^2 + \sum_{i=1}^{m-1} \sum_{j=i+1}^{m} x_i^2 x_j^2 + \cdots + x_1^2 x_2^2 \ldots x_m^2. \tag{1.46}$$

It is clear that this expression is maximal and equal to n in the design points. This is because each x_i equals ± 1 in each design point so that $x_i^2 = 1$. Any other point in the region of interest has at least one coordinate $x_i \neq \pm 1$ so that $x_i^2 < 1$ and $\mathbf{f}'(\mathbf{x})\mathbf{f}(\mathbf{x}) < n$. Therefore, (1.36) achieves its minimum at the design points. Moreover, $p - d(\mathbf{x}, \xi) = 0$ since $p = n$. Conditions 2 and 3 of the General Equivalence Theorem are thus satisfied and the two-level factorial design is $\mathcal{D}$- as well as $\mathcal{G}$-optimal. The design remains optimal if one or more main effects and/or interaction effects are not included in the model.

The class of 2^m factorials can readily be divided into $b = 2^q$ blocks of size 2^{m-q}. This is done at the expense of information on higher order interactions. An important feature of the blocked 2^m factorials is that the model parameters can be estimated without any effect from the blocking. Blocking two-level factorial is thoroughly described in Chapter 9. In this chapter, the specific notation used in the context of factorial experiments is explained in detail. We show in Section 2.3.6 that the blocked designs obtained in this way are $\mathcal{D}$-optimal.

2^{m-f} fractional factorial designs

A disadvantage of the 2^m factorials is that the number of observations needed increases rapidly with the number of experimental factors m. If the higher order interactions are assumed to be negligible, a 2^{m-f} fractional factorial design can be considered instead. This would imply that only a fraction 2^{-f} of the 2^m factorial has to be carried out. For example, a $2^{4-1} = 2^3$ factorial can be used to investigate the effects of four experimental factors instead of a 2^4 factorial. The quality of this type of experiments is expressed in terms of resolution. The lower the resolution of a fractional factorial, the more the main effects are entangled with low order interactions and the more difficult the interpretation of the results is. For example, in a fractional factorial of resolution III, the main effects and some two factor interactions cannot be distinguished from each other. In a fractional factorial of resolution IV, the main effects are confounded with interactions of order three or higher, but not with two-factor interactions. To distinguish between designs with equal resolution, Fries and Hunter (1980) introduced the minimum aberration criterion, which minimizes the number of low order effects confounded with each other. A thorough treatment of 2^{m-f} fractional factorial designs is given in Chapter 9.

Like the 2^m factorials, the 2^{m-f} fractional factorials are orthogonal designs and the information matrix is a diagonal matrix. The diagonal elements are all equal to $n = 2^{m-f}$. As a result, it can be shown that the 2^{m-f} fractional factorials are both $\mathcal{D}$- and $\mathcal{G}$-optimal for all estimable models for which the full factorial design is optimal. The set of estimable models corresponding to a 2^{m-f} fractional factorial is a subset of the models that can be estimated by using the 2^m factorial design.

Plackett-Burman designs

A third class of orthogonal first order designs for n a multiple of four and for $m = n - 1$, was introduced by Plackett and Burman (1946). Their objective was to obtain screening designs that allow the estimation of all main effects with maximum precision. If n is a power of two, the Plackett-Burman designs are identical to the standard fractional factorial designs. For models containing main effects only, the information matrix is diagonal with diagonal elements n and the Plackett-Burman designs are $\mathcal{D}$- and $\mathcal{G}$-optimal.

To construct a Plackett-Burman design, a row of m + and $-$ signs is selected so that the number of positive signs is equal to $(m + 1)/2$ and the number of negative signs is equal to $(m - 1)/2$. Note that both $m + 1$ and $m - 1$ are divisible by two since $m + 1$ is a multiple of four. This row is chosen as the first row in the design. The second row is obtained from the first by shifting it one place, and so forth. Finally, a row of $-$ signs is

Table 1.6: 8-point Plackett-Burman design for $m = 7$.

	x_1	x_2	x_3	x_4	x_5	x_6	x_7
1	+1	+1	+1	-1	+1	-1	-1
2	-1	+1	+1	+1	-1	+1	-1
3	-1	-1	+1	+1	+1	-1	+1
4	+1	-1	-1	+1	+1	+1	-1
5	-1	+1	-1	-1	+1	+1	+1
6	+1	-1	+1	-1	-1	+1	+1
7	+1	+1	-1	+1	-1	-1	+1
8	-1	-1	-1	-1	-1	-1	-1

added to the previous rows, producing a design with n rows. For example, for $m = 7$, the generator

$$+1 \ +1 \ +1 \ -1 \ +1 \ -1 \ -1$$

leads to the design in Table 1.6. It should be pointed out that the Plackett-Burman designs remain $\mathcal{D}$- and $\mathcal{G}$-optimal when $m < n - 1$ as well.

Simplex design

The simplex design is an orthogonal design with $n = m + 1$ design points and was introduced by Box (1952). The design points are located at the vertices of an m-dimensional simplex. They are characterized by the fact that the angle θ, which any two points make with the origin is so that $\cos(\theta) = -1/m$. For $m = 2$, the design points are the three vertices of an equilateral triangle. Unlike the other first order designs described here, the simplex design can only be used with quantitative factors.

1.5.2   Second order designs

In contrast with first order designs, second order designs need at least three levels for each factor under investigation. Otherwise, the quadratic effects cannot be estimated and the information matrix (or the corresponding design) is singular. In this section, we will present several types of designs for the estimation of model (1.40).

3^m factorial designs

One possible second-order design is the 3^m factorial design, consisting of all combinations of points at which the factors take coded values of -1, 0 or $+1$. As an illustration, the 3^3 factorial is displayed in Figure 1.5. It is clear that the number of observations needed increases rapidly as m increases. For larger m, the 3^{m-f} fractional factorials can be considered. Blocking 3^m factorials and constructing 3^{m-f} fractional factorials receives attention

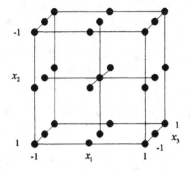

Figure 1.5: 3^3 factorial design.

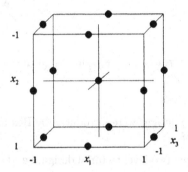

Figure 1.6: Box-Behnken design for $m = 3$.

in Montgomery (1991) and Wu and Hamada (2000). In contrast with their two-level counterparts, three-level factorials are in general not $\mathcal{D}$-optimal for estimating a second order model. They perform, however, well with respect to the $\mathcal{A}$-optimality criterion.

Box-Behnken designs

Another class of three-level designs for estimating the parameters of a second order model was developed by Box and Behnken (1960). The Box-Behnken designs are formed by combining a balanced incomplete block design (see Section 1.6) and a two-level factorial design. The three-variable Box-Behnken design possesses 13 design points, one of which is the center point. It is graphically displayed in Figure 1.6. Apart from the center point, all design points lie on a distance $\sqrt{2}$ from the center point. Therefore, the Box-Behnken designs are especially useful when the design region is hyperspherical. When the design region is hypercubic, it is an inefficient design option.

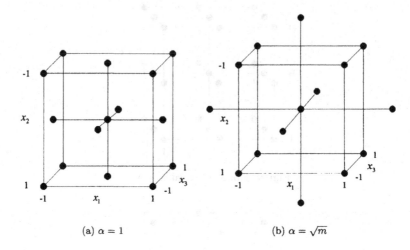

(a) $\alpha = 1$ (b) $\alpha = \sqrt{m}$

Figure 1.7: Central composite designs for $m = 3$.

Central composite designs

The central composite designs were introduced by Box and Wilson (1951). The designs consist of

1. a full or fractional two-level factorial design, i.e. the factorial portion,

2. at least one observation at the center point,

3. two axial points on the axis of each design variable at a distance α from the center point.

When the design region is hypercubic, α is set equal to one. The resulting design is a face centered central composite design. When the design region is hyperspherical, α is usually set equal to $\sqrt{m}$. In Figure 1.7, both a face centered and an ordinary central composite design for $m = 3$ are displayed. If $\alpha = \sqrt{m}$, all design points except the center point itself lie on a distance $\sqrt{m}$ from the center point. Hartley (1959) investigated what fractional factorial designs should be used in a central composite design in order to permit the estimation of the largest possible number of coefficients. Lucas (1974, 1976, 1977) found out that the $\mathcal{D}$-efficiency of the face centered central composite design is quite high for $m \leq 5$. For an ordinary central composite design, it is high for all values of m considered. Finally, it can be shown that the $\mathcal{D}$-criterion value is an increasing function of α. For more information on properties like orthogonality, rotatability or uniform precision of central composite designs, we refer the reader to Khuri and Cornell (1987) and the references therein.

Other second order designs

Many other designs for estimating a second order model have been proposed. Equiradial designs are designs consisting of two or more sets of points, where the points in each set are equidistant from the origin. An example of an equiradial design is the central composite design in Figure 1.7b. Box and Draper (1971) proposed a class of saturated second order designs, where the design points were chosen to maximize $|\mathbf{X'X}|$. Hoke (1974) presented economical second order designs based on irregular fractions of the 3^k factorial. Roquemore (1976) introduced the hybrid designs, which are based on the central composite designs. An important drawback of the hybrid designs is their multitude of factor levels. Draper and Lin (1990) derive a number of small designs of composite type by using the columns of the Plackett-Burman designs instead of the two-level factorial or fractional factorial designs. Mee (2001) constructs noncentral composite designs by combining two first order designs with different centers.

1.6 Categorical designs

In the literature, the representation of categorical and response surface designs is completely different. Response surface designs or regression designs are usually given by their design matrix $\mathbf{X}$ or by a geometrical representation. For categorical designs, the same information is summarized in tabular form. In addition, the mathematical models corresponding to categorical designs look totally different and the terminology is totally different. Finally, experiments with categorical variables are mostly analyzed by using analysis of variance tables. Many of the designs described here are described by Montgomery (1991). The definitions and properties of the designs are given by Shah and Sinha (1989). Catalogues of categorical designs can be found in Cochran and Cox (1957) and Cox (1958).

1.6.1 Completely randomized design

When there is no heterogeneity in the experimental units utilized for treatment comparisons, a completely randomized design is used. The statistical model is written by

$$y_{ij} = \mu + \tau_i + \varepsilon_{ij}, \tag{1.47}$$

where y_{ij} is the response obtained at the jth replication of the ith treatment, μ is the overall mean, τ_i represents the ith treatment effect and ε_{ij} is the random error. In this type of experiment, treatments are allotted to the experimental units at random. When all experimental factors have two levels, the 2^m factorial might be a design option. Similarly, the 3^m factorial can be considered if all factors have three levels. When the number of factor levels is not the same for all factors, a mixed level factorial can be used.

The completely randomized design has the advantage of flexibility when compared to the block designs described in the sequel of this section. As a matter of fact, for certain combinations of block size, number of treatments and number of blocks, no suitable block designs exists. In addition, the categorical block designs described in the literature assume that all treatments are replicated the same number of times. However, a completely randomized design is not the best design option when the experimental units are not homogeneous. Of course, the source of variation causing this heterogeneity should be taken into account when designing and analyzing the experiment. The resulting experiment is a blocked experiment and the experimental units within each block are as homogeneous as possible. Block effects are included in the statistical model in order to avoid a less powerful analysis due to an inflation of the random errors. It should be noted at this point that blocking experiments is not exclusively a matter of categorical variables. Blocking experiments with continuous variables is the topic of Section 2.3 and Chapters 4 and 5.

1.6.2   One blocking variable

The statistical model for treatment comparison in the presence of one blocking factor is given by

$$y_{ij} = \mu + \tau_i + \alpha_j + \varepsilon_{ij}, \tag{1.48}$$

where μ is the overall mean, τ_i is the ith treatment effect, α_j is the jth block effect and ε_{ij} is the error term. Appropriate designs for the estimation of this model are balanced block designs and group divisible designs. Blocked experiments are typically used when not all the runs can be performed on one day, with one batch of material or on one field.

Balanced block designs

An important design for the comparison of t treatments in the presence of one blocking factor is the balanced block design. A simple definition is given in Cox (1958):

1. each block contains the same number of observations k,

2. each treatment occurs the same number of times r in the entire experiment,

3. the number of times two different treatments occur together in a block is equal for all pairs of treatments.

If we denote the number of blocks by b, the total number of observations n is equal to $bk = rt$. When the block size is equal to the number of different treatments in the design, the balanced block design is said to be complete. In Neter, Kutner, Nachtsheim and Wasserman (1996), balanced complete

Table 1.7: Balanced incomplete block design with five treatments and ten blocks of size three.

Block	Treatments			Block	Treatments		
1	1	2	3	6	1	4	5
2	1	2	4	7	2	3	4
3	1	2	5	8	2	3	5
4	1	3	4	9	2	4	5
5	1	3	5	10	3	4	5

Table 1.8: 3 × 3 lattice design.

Block	Replicate I			Block	Replicate III		
1	1	2	3	7	1	5	9
2	4	5	6	8	2	6	7
3	7	8	9	9	3	4	8
Block	Replicate II			Block	Replicate IV		
4	1	4	7	10	1	6	8
5	2	5	8	11	2	4	9
6	3	6	9	12	3	5	7

block designs are referred to as randomized complete block designs. When the block size is smaller than the number of different treatments, the balanced block design reduces to a balanced incomplete block design. Finally, when the number of treatments t is equal to the number of blocks b, the design is said to be symmetric. Table 1.7 shows an example of a balanced incomplete block design with ten blocks of size three for the comparison of five treatments. When a balanced block design is used, the treatment effects are estimated with the highest possible precision given the number of observations n, the number of blocks b and the homogeneous block size k. It also turns out that the class of balanced block designs is universally optimal for the estimation of model (1.48), i.e. a balanced block design is $\mathcal{A}$-, $\mathcal{D}$- and $\mathcal{E}$-optimal as well as optimal with respect to any generalized optimality criterion. In other words, no better design with n observations and b blocks of size k can be found to compare t treatments. This result is also valid when the blocks are random instead of fixed.

A special case of a balanced incomplete block design is a lattice design. When the number of treatments is the square of the block size k and each treatment is replicated $k + 1$ times in the entire experiment, the blocks can be divided in $k + 1$ groups so that each treatment occurs once in each group. The 3 × 3 lattice design is displayed in Table 1.8.

Table 1.9: Group divisible design with six blocks of size three for the comparison of six treatments.

Block	Treatments		
1	1	2	3
2	1	2	4
3	1	5	6
4	2	5	6
5	3	4	5
6	3	4	6

Group divisible designs

Unfortunately, practical balanced incomplete block designs do not exist for all combinations of block size and number of treatments. For example, a balanced incomplete block design for eight treatments and block size three would require 56 blocks. Each treatment would then be replicated 21 times. In such cases, group divisible or partially balanced incomplete block designs may provide the experimenter with a smaller design. In a group divisible design, the number of times two different treatments occur together in the same block is no longer the same for all pairs of treatments. A formal definition is given in Montgomery (1991) and in Shah and Sinha (1989). In order to construct a group divisible design, the treatments are divided in groups. Two treatments belonging to the same group then occur together in k_1 blocks, while two treatments belonging to different groups occur together k_2 times. When $k_2 = k_1 \pm 1$, the design is said to be most balanced. The optimality of this type of design is discussed in Cheng (1978). When $k_1 = k_2$, we obtain a balanced block design. In Table 1.9, a group divisible design with six blocks of size three for the comparison of six treatments is shown. A detailed overview of the statistical properties of this type of design is given in Shah and Sinha (1989).

1.6.3   Two blocking variables

Categorical designs that take into account two blocking factors are often referred to as row-column designs. The corresponding statistical model is given by

$$y_{ijk} = \mu + \tau_i + \alpha_j + \beta_k + \varepsilon_{ijk}, \qquad (1.49)$$

where μ is the overall mean, τ_i is the ith treatment effect, α_j is the jth row effect, β_k is the kth column effect and ε_{ijk} is the error term. One design that can be used to eliminate the effect of the two blocking variables is the generalized Youden design. The main feature of generalized Youden designs is that they form a balanced block design for both blocking variables. When one of the blocking variables has as many levels as there are treatments, the

Table 1.10: Youden square design for seven treatments with four rows and seven columns.

	Column						
Row	1	2	3	4	5	6	7
1	A	B	C	D	E	F	G
2	C	D	E	F	G	A	B
3	D	E	F	G	A	B	C
4	E	F	G	A	B	C	D

Table 1.11: Latin square design for seven treatments.

	Column						
Row	1	2	3	4	5	6	7
1	A	B	C	D	E	F	G
2	B	C	D	E	F	G	A
3	C	D	E	F	G	A	B
4	D	E	F	G	A	B	C
5	E	F	G	A	B	C	D
6	F	G	A	B	C	D	E
7	G	A	B	C	D	E	F

design is a Youden square design. An example is given in Table 1.10. When both blocking variables have as many levels as there are treatments, the design is a Latin square design. A Latin square design for seven treatments is given in Table 1.11. Note that the Youden square design in Table 1.10 consists of the first, third, fourth and fifth row of the Latin square design in Table 1.11. Shah and Sinha (1989) show that the Youden square designs as well as the Latin square designs are universally optimal when the blocks are fixed as well as when they are random.

Montgomery (1991) describes a chemical experiment in which a Latin square design was used. The purpose of the experiment was to investigate the effect of five ingredients on the reaction time of a chemical process. Each batch of new material was only large enough to permit five runs to be made. Furthermore, only five runs could be made in one day. As a result, five different batches of material were used and the experiment was run on five different days. Therefore, there are two controllable sources of variation.

1.6.4   Three blocking variables

A Graeco-Latin square design can be used to control three sources of extraneous variability, i.e. to block in three directions. A Graeco-Latin square design is obtained by superimposing two different Latin square designs. If

Table 1.12: 4 × 4 Graeco-Latin square design.

Row	Column			
	1	2	3	4
1	Aα	Bβ	Cγ	Dδ
2	Bδ	Aγ	Dβ	Cα
3	Cβ	Dα	Aδ	Bγ
4	Dγ	Cδ	Bα	Aβ

the treatments of the first and second Latin square design are denoted by Latin and Greek letters respectively, then each Greek letter should appear only once with each Latin letter. Under this notation, the Greek letters represent the levels of the third blocking factor. A 4 × 4 Graeco-Latin square design is shown in Table 1.12. The corresponding statistical model is given by

$$y_{ijkl} = \mu + \tau_i + \alpha_j + \beta_k + \gamma_l + \varepsilon_{ijkl}, \qquad (1.50)$$

where μ is the overall mean, τ_i is the ith treatment effect, α_j is the jth row effect, β_k is the kth column effect, γ_l is the effect of the lth Greek letter and ε_{ijkl} is the error term.

An example of an experiment in which a Graeco-Latin square design is used can be found in Davies (1945). The purpose of the experiment was to investigate the effect of seven types of gasoline on the miles per gallon achieved by a particular car. In the experiment, there were three potential sources of variation influencing the results: the driver of the car, the day of the test, and the time of the day. Therefore, seven drivers, seven days and seven different times were used in a Graeco-Latin square.

1.6.5   Cross-over designs

In many fields of scientific investigation, experiments need to be designed in such a manner that each experimental unit or subject receives some or all of the treatments over a certain period of time. Such designs have been discussed in the literature under names as cross-over or change-over designs, time series designs or repeated measurement designs. A repeated measurement design can be viewed as a row-column design with a set of subjects displayed across the columns and a set of periods displayed across the rows. The peculiarity of this type of experiment is that any treatment applied to a unit in a certain period not only influences the response in that period, but it may also influence the response in the next periods. Mostly, it is assumed that only the carry-over effect to the next period is of importance. The statistical model can then be written as

$$y_{ij} = \mu + \tau_{d(i,j)} + \rho_{d(i,j-1)} + \alpha_i + \pi_j + \varepsilon_{ij}, \qquad (1.51)$$

where the design allocates treatment $d(i, j)$ to the ith subject in the period j, the treatment and the carry-over effect are represented by $\tau_{d(i,j)}$ and $\rho_{d(i,j-1)}$ respectively, α_i is the ith subject effect, π_j is the effect of period j and ε_{ij} is the random error. Shah and Sinha (1989) show that some specific types of cross-over designs are universally optimal. An algorithm to construct designs that allow efficient pairwise comparisons is presented by Donev (1997).

1.7   The $\mathcal{D}$-optimality criterion

For our computations, we have chosen the $\mathcal{D}$-optimality criterion for three reasons. Firstly, $\mathcal{D}$-optimal designs usually perform well with respect to other design criteria, while the opposite is often not true. Secondly, the $\mathcal{D}$-optimality criterion is invariant to a linear transformation of the design matrix. As a result, it is invariant to the scale or the coding of the variables. Finally, the $\mathcal{D}$-optimality criterion has the advantage of computational simplicity thanks to the existence of powerful update formulae for the determinant and the inverse of the information matrix. In this section, we will deal with the first two reasons in some more detail. The third reason will receive attention in the next section.

It turns out that, in contrast with other design criteria, the $\mathcal{D}$-optimality criterion produces designs that are efficient with respect to other criteria as well. For instance, Lucas (1974) points out that the best design in terms of $\mathcal{D}$-efficiency often coincides with the best design in terms of $\mathcal{G}$-efficiency. Donev and Atkinson (1988) show that $\mathcal{D}$-optimal response surface designs possess high $\mathcal{G}$- and $\mathcal{V}$-efficiencies as well. Chasalow (1992) evaluated a collection of designs with respect to both $\mathcal{D}$- and $\mathcal{A}$-optimality and discovered that designs which perform well with respect to $\mathcal{A}$-optimality often perform poorly with respect to $\mathcal{D}$-optimality. In contrast, the best designs with respect to $\mathcal{D}$-optimality tend to perform well with respect to $\mathcal{A}$-optimality too.

Unlike other design criteria, the $\mathcal{D}$-optimality criterion is invariant to a linear transformation of the design matrix $\mathbf{X}$. Letting $\mathbf{Z} = \mathbf{XA}$ with $\mathbf{A}$ a $p \times p$ matrix that does not depend on the design, the $\mathcal{D}$-criterion value becomes

$$
\begin{aligned}
|\mathbf{Z}'\mathbf{Z}| &= |(\mathbf{XA})'\mathbf{XA}|, \\
&= |\mathbf{A}'\mathbf{X}'\mathbf{XA}|, \\
&= |\mathbf{A}'||\mathbf{X}'\mathbf{X}||\mathbf{A}|, \\
&= |\mathbf{A}|^2|\mathbf{X}'\mathbf{X}|.
\end{aligned}
$$

The $\mathcal{D}$-criterion value of the transformed design matrix is thus proportional to that of the original design matrix. For other design criteria, this result does not hold. In practice, this implies that the ordering of designs with respect to the $\mathcal{D}$-optimality criterion is independent of the coding of the variables. On the contrary, the ordering depends on the coding when the $\mathcal{A}$-optimality criterion is used.

Consider an experiment in which the effect of four treatments on a certain response is to be investigated. Although the statistical model is very simple, it can be written in at least three ways. A first representation is

$$y_{ij} = \sum_{i=1}^{4} \alpha_i x_{1i} + \varepsilon_{ij}, \tag{1.52}$$

where $\alpha = [\, \alpha_1 \; \alpha_2 \; \alpha_3 \; \alpha_4 \,]'$ is the vector containing the mean responses for the treatments and x_{1i} $(i = 1, 2, 3, 4)$ is equal to one if the ith treatment is given and zero otherwise. A second representation is given by

$$y_{ij} = \sum_{i=1}^{3} \beta_i x_{2i} + \beta_4 + \varepsilon_{ij}, \tag{1.53}$$

where $\beta = [\, \beta_1 \; \beta_2 \; \beta_3 \; \beta_4 \,]'$ with $\beta_4 = \alpha_4$, $\beta_1 = \alpha_1 - \alpha_4$, $\beta_2 = \alpha_2 - \alpha_4$ and $\beta_3 = \alpha_3 - \alpha_4$, and x_{2i} $(i = 1, 2, 3)$ is equal to one if the ith treatment is given and zero otherwise. This notation is especially useful if the fourth treatment is a control treatment. Finally, a third reparameterization can be written as

$$y_{ij} = \gamma_0 + \sum_{i=1}^{3} \gamma_i x_{3i} + \varepsilon_{ij}, \tag{1.54}$$

where γ_0 represents the overall mean, $\gamma_i = \alpha_i - \gamma_0$ $(i = 1, 2, 3)$, and x_{3i} equals 1 if the ith treatment is given, -1 if the last treatment is given and zero otherwise. The way the model is represented influences the design matrix. As an illustration, consider an experiment with 36 observations in which each treatment is replicated nine times. If we denote by $\mathbf{a}_k$ a k-dimensional vector of a's, the design matrix can be written as

$$\mathbf{X}_1 = \begin{bmatrix} \mathbf{1}_9 & \mathbf{0}_9 & \mathbf{0}_9 & \mathbf{0}_9 \\ \mathbf{0}_9 & \mathbf{1}_9 & \mathbf{0}_9 & \mathbf{0}_9 \\ \mathbf{0}_9 & \mathbf{0}_9 & \mathbf{1}_9 & \mathbf{0}_9 \\ \mathbf{0}_9 & \mathbf{0}_9 & \mathbf{0}_9 & \mathbf{1}_9 \end{bmatrix},$$

if the first model is used,

$$\mathbf{X}_2 = \begin{bmatrix} \mathbf{1}_9 & \mathbf{0}_9 & \mathbf{0}_9 & \mathbf{1}_9 \\ \mathbf{0}_9 & \mathbf{1}_9 & \mathbf{0}_9 & \mathbf{1}_9 \\ \mathbf{0}_9 & \mathbf{0}_9 & \mathbf{1}_9 & \mathbf{1}_9 \\ \mathbf{0}_9 & \mathbf{0}_9 & \mathbf{0}_9 & \mathbf{1}_9 \end{bmatrix},$$

if the second model is chosen, and

$$\mathbf{X}_3 = \begin{bmatrix} 1_9 & 1_9 & 0_9 & 0_9 \\ 1_9 & 0_9 & 1_9 & 0_9 \\ 1_9 & 0_9 & 0_9 & 1_9 \\ 1_9 & -1_9 & -1_9 & -1_9 \end{bmatrix},$$

if the third model is used. The $\mathcal{D}$-criterion values corresponding to $\mathbf{X}_1$, $\mathbf{X}_2$ and $\mathbf{X}_3$ amount to 6561, 6561 and 104976 respectively. The $\mathcal{A}$-criterion values amount to 0.44, 0.78 and 0.28 respectively. Since postmultiplying $\mathbf{X}_1$ by

$$\mathbf{A}_{12} = \begin{bmatrix} 1 & 0 & 0 & 1 \\ 0 & 1 & 0 & 1 \\ 0 & 0 & 1 & 1 \\ 0 & 0 & 0 & 1 \end{bmatrix}$$

yields $\mathbf{X}_2$, and postmultiplying $\mathbf{X}_2$ by

$$\mathbf{A}_{23} = \begin{bmatrix} 0 & 2 & 1 & 1 \\ 0 & 1 & 2 & 1 \\ 0 & 1 & 1 & 2 \\ 1 & -1 & -1 & -1 \end{bmatrix}$$

yields $\mathbf{X}_3$, it is clear that $\mathbf{X}_1$, $\mathbf{X}_2$ and $\mathbf{X}_3$ are linear transformations of each other and that the model used does not affect the $\mathcal{D}$-efficiency of the design. This is further demonstrated by comparing the experiment with nine replications of each treatment to an experiment in which the first three treatments are replicated ten times and the fourth is replicated six times. The second design may be inspired by the fact that the fourth treatment is more expensive than the others or by the fact that it is less important. Its design matrix is given by

$$\mathbf{Z}_1 = \begin{bmatrix} 1_{10} & 0_{10} & 0_{10} & 0_{10} \\ 0_{10} & 1_{10} & 0_{10} & 0_{10} \\ 0_{10} & 0_{10} & 1_{10} & 0_{10} \\ 0_6 & 0_6 & 0_6 & 1_6 \end{bmatrix},$$

if the first model is used,

$$\mathbf{Z}_2 = \begin{bmatrix} 1_{10} & 0_{10} & 0_{10} & 1_{10} \\ 0_{10} & 1_{10} & 0_{10} & 1_{10} \\ 0_{10} & 0_{10} & 1_{10} & 1_{10} \\ 0_6 & 0_6 & 0_6 & 1_6 \end{bmatrix},$$

if the second model is chosen, and

$$\mathbf{Z}_3 = \begin{bmatrix} 1_{10} & 1_{10} & 0_{10} & 0_{10} \\ 1_{10} & 0_{10} & 1_{10} & 0_{10} \\ 1_{10} & 0_{10} & 0_{10} & 1_{10} \\ 1_6 & -1_6 & -1_6 & -1_6 \end{bmatrix},$$

if the third model is used. The $\mathcal{D}$-criterion values corresponding to $\mathbf{Z}_1$, $\mathbf{Z}_2$ and $\mathbf{Z}_3$ amount to 6000, 6000 and 96000 respectively. The $\mathcal{A}$-criterion values amount to 0.47, 0.97 and 0.27 respectively. From these results, it can be seen that, independent of the parameterization used, the first design is 2.26% more efficient than the second in terms of $\mathcal{D}$-efficiency. On the contrary, the relative $\mathcal{A}$-efficiency and even the ordering of both designs depends on the model. If the first or second model are chosen, then the first design has a better $\mathcal{A}$-criterion value than the second. However, if the third model is used, the second design should be preferred in terms of $\mathcal{A}$-optimality.

The $\mathcal{D}$-optimality criterion is also invariant to a linear transformation of the factor levels. This is a special case of linearly transforming the design matrix. Consider for example quadratic regression on one variable x and suppose that a linear transformation of the factor levels $z = b(a + x)$ is carried out. It can be verified that any row

$$\begin{bmatrix} 1 & z & z^2 \end{bmatrix} = \begin{bmatrix} 1 & b(a+x) & b^2(a+x)^2 \end{bmatrix}$$

of the transformed design matrix is equal to

$$\begin{bmatrix} 1 & x & x^2 \end{bmatrix} \begin{bmatrix} 1 & ba & b^2a^2 \\ 0 & b & 2b^2a \\ 0 & 0 & b^2 \end{bmatrix}.$$

A final advantage of the $\mathcal{D}$-optimality criterion is that the effect of a design change can be computed at a low computational cost. This point is extremely important for design construction algorithms because the computation of optimal designs requires the evaluation of many design changes. In the next section, we show how the information matrix on the unknown β and the variance-covariance matrix of $\hat{\beta}$ can be updated after the addition or the deletion of a design point, and after the substitution of a design point. We also show how the $\mathcal{D}$-criterion value can be updated after these changes.

1.8   Updating the information matrix, its inverse and its determinant

The low computational cost of updating the information matrix, its inverse and its determinant after a design change is a direct consequence of the fact that the information matrix can be written as a sum of outer products:

$$\mathbf{X}'\mathbf{X} = \sum_{i=1}^{n} \mathbf{f}(\mathbf{x}_i)\mathbf{f}'(\mathbf{x}_i). \tag{1.55}$$

In this expression $f(x_i)$ is the ith row of X, or alternatively, the polynomial expansion of the ith design point. In order to update the information matrix after the addition or the deletion of an observation in design point a, the outer product $f(a)f'(a)$ is added to or subtracted from (1.55):

$$X^{*'}X^* = X'X \pm f(a)f'(a).$$

As a consequence of Theorem 18.1.1 of Harville (1997), the modified $\mathcal{D}$-criterion value can be written as

$$|X'X \pm f(a)f'(a)| = |X'X|\{1 \pm f'(a)(X'X)^{-1}f(a)\}. \qquad (1.56)$$

From this expression, it can be seen that the greatest improvement of the $\mathcal{D}$-criterion value can be generated by adding the point with the largest prediction variance $f'(a)(X'X)^{-1}f(a)$ to the design. Similarly, if a point has to be deleted from the design, the point with the smallest prediction variance should be removed. Exchanging point b by point a modifies the $\mathcal{D}$-criterion value by a factor

$$\{1 + f'(a)(X'X)^{-1}f(a)\}\{1 - f'(b)(X'X)^{-1}f(b)\} + \{f'(a)(X'X)^{-1}f(b)\}^2. \qquad (1.57)$$

It is probable that this expression will be large if a point with a small prediction variance is replaced by a point with a large prediction variance. For this reason, Atkinson and Donev (1989) only consider points with a large prediction variance as candidates for entering the design and points with a small prediction variance as candidates to be deleted from the design. As a consequence of Corollary 18.2.10 of Harville (1997), the inverse of the information matrix after addition or deletion of a design point can be calculated as

$$\{X'X \pm f(a)f'(a)\}^{-1} = (X'X)^{-1} \mp \frac{\{(X'X)^{-1}f(a)\}\{(X'X)^{-1}f(a)\}'}{1 \pm f'(a)(X'X)^{-1}f(a)}. \qquad (1.58)$$

In order to illustrate the application of these update formulae, let us revisit the gas turbine experiment from Section 1.2.2. It can be verified that the $\mathcal{D}$-criterion value, given by the determinant of $X'X$, of the original experiment is equal to 10944. Suppose that one of the center runs, i.e. run 9, 10 or 11, is removed from the design, then the updated information matrix can be calculated as

$$X'X - \begin{bmatrix} 1 \\ 0 \\ 0 \\ 0 \\ 0 \\ 0 \end{bmatrix} \begin{bmatrix} 1 & 0 & 0 & 0 & 0 & 0 \end{bmatrix},$$

yielding

$$\mathbf{X^{*'}X^*} = \begin{bmatrix} 10 & 0 & 0 & 0 & 6 & 6 \\ 0 & 6 & 0 & 0 & 0 & 0 \\ 0 & 0 & 6 & 0 & 0 & 0 \\ 0 & 0 & 0 & 4 & 0 & 0 \\ 6 & 0 & 0 & 0 & 6 & 4 \\ 6 & 0 & 0 & 0 & 4 & 6 \end{bmatrix}.$$

The determinant of this matrix equals

$$|\mathbf{X^{*'}X^*}| = |\mathbf{X'X}| \{ 1 - \begin{bmatrix} 1 & 0 & \cdots & 0 \end{bmatrix} (\mathbf{X'X})^{-1} \begin{bmatrix} 1 \\ 0 \\ \vdots \\ 0 \end{bmatrix} \},$$

$$= 10944 \times (1 - 0.2632),$$
$$= 8064.$$

As a result, the $\mathcal{D}$-criterion value of the experiment is much smaller after deleting the center point. In a similar fashion, the effect of deleting other runs or design points can be verified. For example, deleting the first run yields a determinant of 2256. As a result, the first run is much more important for the quality of the experiment than one of the center runs. This is mainly due to the fact that the center run is replicated three times in the original gas turbine experiment. Since

$$\begin{bmatrix} 1 & 0 & 0 & 0 & 0 & 0 \end{bmatrix} (\mathbf{X'X})^{-1} = \begin{bmatrix} 0.2632 & 0 & 0 & 0 & -0.1579 & -0.1579 \end{bmatrix},$$

the inverse of the information matrix becomes

$$(\mathbf{X^{*'}X^*})^{-1} = (\mathbf{X'X})^{-1} + (1 - 0.2632)^{-1} \begin{bmatrix} 0.2632 \\ 0 \\ 0 \\ 0 \\ -0.1579 \\ -0.1579 \end{bmatrix} \begin{bmatrix} 0.2632 \\ 0 \\ 0 \\ 0 \\ -0.1579 \\ -0.1579 \end{bmatrix}',$$

$$= (\mathbf{X'X})^{-1} + 1.3572 \begin{bmatrix} 0.0693 & 0 & 0 & 0 & -0.0416 & -0.0416 \\ 0 & 0 & 0 & 0 & 0 & 0 \\ 0 & 0 & 0 & 0 & 0 & 0 \\ 0 & 0 & 0 & 0 & 0 & 0 \\ -0.0416 & 0 & 0 & 0 & 0.0249 & 0.0249 \\ -0.0416 & 0 & 0 & 0 & 0.0249 & 0.0249 \end{bmatrix},$$

$$
= \begin{bmatrix}
0.3571 & 0 & 0 & 0 & -0.2143 & -0.2143 \\
0 & 0.1667 & 0 & 0 & 0 & 0 \\
0 & 0 & 0.1667 & 0 & 0 & 0 \\
0 & 0 & 0 & 0.25 & 0 & 0 \\
-0.2143 & 0 & 0 & 0 & 0.4286 & -0.0714 \\
-0.2143 & 0 & 0 & 0 & -0.0714 & 0.4286
\end{bmatrix}.
$$

It is clear that evaluating design changes and updating the information matrix and its inverse is simplified by the equations given in this section. They also allow a fast update of the $\mathcal{D}$-criterion value after design changes. For the sake of accuracy, it is, however, useful to recompute the information matrix directly from (1.10) after a number of design changes. In the next section, we give a brief overview of the existing algorithms for the construction of $\mathcal{D}$-optimal designs. Of course, the fast update formulae for the information matrix, its inverse and its determinant are intensively used in these algorithms.

1.9   Constructing discrete $\mathcal{D}$-optimal designs

A vast literature on the construction of exact, tailor-made $\mathcal{D}$-optimal response surface designs under the assumption of independent and identically distributed errors with zero mean and homoscedastic variance (see (1.4)-(1.7)) can be found. Many of the early approaches involved the direct maximization of $|\mathbf{X'X}|$ by mathematical programming techniques. Because of dimensionality problems, attention has shifted towards a number of methods which take into account the special characteristics of the design problem.

The ability to compute continuous $\mathcal{D}$-optimal designs (see Fedorov (1972)) led Kiefer (1971) to suggest simply rounding off the continuous design to obtain the $\mathcal{D}$-optimal exact design. He added that for large n, exact designs obtained in this way are nearly $\mathcal{D}$-optimal. Welch (1982) uses a branch-and-bound algorithm to find the globally best exact design for a given design criterion and set of candidate points. However, the computational complexity is such that even moderate problems require appreciable computation time. Haines (1987) and Meyer and Nachtsheim (1988) used simulated annealing to construct $\mathcal{D}$-optimal designs for continuous design spaces.

The bulk of the remaining approaches may be classified as exchange algorithms. These procedures start with the construction of a non-singular n-point design. Next, they add and delete one or more observations in order to achieve increases in the determinant $|\mathbf{X'X}|$. Design points are chosen from a predefined set of candidate or support points that cover the entire design region. In order to avoid being stuck in a local optimum, most algorithms repeat the search for a number of different starting designs. Each repetition is called a try. The generation of starting designs often includes a random component after which the starting design is completed by using a greedy heuristic to select design points from the set of candidates. Each time it is called, the greedy heuristic picks out the point that produces the largest increase in the determinant and adds it to the starting design.

The first exchange algorithm for the computation of exact $\mathcal{D}$-optimal designs was developed by Fedorov (1972). An important feature of Fedorov's algorithm is that the points to be exchanged are chosen simultaneously to maximize (1.57). It is therefore said to be a rank 2 algorithm. Wynn (1972) proposed a modified version in which the added and the deleted design point are determined in two separate steps. Therefore, his algorithm is of rank 1. By splitting the exchange in two steps, the candidate point with the largest prediction variance will be added to the design, while the design point with the smallest prediction variance will be removed. Mitchell (1974a) proposed the rank 1 DETMAX algorithm and allows the design size to vary between $n-k$ and $n+k$ during the search. Cook and Nachtsheim (1980) recommend $k = 4$ for convex design spaces. They also found out that Fedorov's algorithm often produces better designs, but requires more computation time than its competitors. The DETMAX algorithm is reasonable in terms of both computation time and $\mathcal{D}$-efficiency. It also turns out that the worse the starting design, the longer it takes the algorithm to converge to the local optimum. Similar conclusions were drawn in Johnson and Nachtsheim (1983) and Nguyen and Miller (1992). Galil and Kiefer (1980) propose several measures to save computation time, the most important of which is a technique for the construction of a starting design. It turns out, however, that using this starting design only leads to a better design in a limited number of cases.

With the BLKL algorithm, Atkinson and Donev (1989, 1992) demonstrated that exchange algorithms are suited for tackling design problems with both quantitative and qualitative variables. In addition, they show that the computation time of rank 2 exchange algorithms can be reduced by considering only the l points with the largest prediction variance as candidates for entering the design and the k points with the smallest prediction variance as candidates to be deleted from the design. Therefore, the algorithm does not become prohibitively slow as the problem size increases in contrast with Fedorov's (1972) algorithm. The logic of this approach, where k and l are user-defined, is described in the previous section. When $k = n$ and l equals the number of candidate points, the BLKL algorithm coincides with Fedorov's (1972) algorithm. Atkinson and Donev (1989) also show that the number of observations at the different levels of the qualitative factors is heterogeneous. This result is counterintuitive because it pleads for unbalanced designs. In addition, ignoring the qualitative factors does not yield a $\mathcal{D}$-optimal design for the case where all factors are quantitative. When a blocked experiment should be constructed, Cook and Nachtsheim (1989) provide an alternative to the BLKL algorithm. In their algorithm, they also consider interchanging design points from different blocks in order to improve the initial design.

Currently, all commercially available software systems use exchange algorithms. The main advantage of this type of algorithms is that many design problems are reduced to a common optimization structure. Whether the design problems involve quantitative or qualitative factors or mixture variables, an optimal design is found by selecting design points from a list of candidate points. The algorithm can therefore handle any form of design region. In most practical applications, a coarse grid on the experimental region of the quantitative variables is used to construct the set of candidates. Although using a finer grid in many cases yields better designs, it is seldom done in practical design problems. A first reason is that the efficiency gain is often negligible. A second reason is that experimenters usually do not like the use of many different factor levels. Finally, computation time rapidly increases as the number of factor levels considered grows larger. Donev and Atkinson (1988) propose to compute an efficient design using a coarse grid on the design region and improving the resulting design by moving away its design points from the original grid. In cases where the design region is constrained, it may be difficult to construct a set of candidate points. In that case, the CONVRT and CONAEV programs of Piepel (1988) and the approach of Goos and Vandebroek (2001b) may help the experimenter to generate the extreme vertices and the various centroids of the experimental region. The vertices and the centroids can then be used as candidate points. A different type of exchange algorithm is the coordinate-exchange algorithm of Meyer and Nachtsheim (1995). Instead of exchanging design points, they consider the exchange of only one factor level at a time. Firstly, the k design points with the smallest prediction variance are identified. Next, it is investigated for each coordinate whether replacing it produces a better design. This method produces highly efficient designs at a low computational cost and does not require the construction of a candidate set. However, this approach is not applicable for mixture experiments and becomes less attractive if subsets of the factors are involved in constraints on the design region.

It should be noted that each of these algorithms can be adapted to generate designs that are optimal with respect to other design criteria. For example, Welch (1984) modified the DETMAX algorithm of Mitchell (1974a) to compute $\mathcal{G}$- and $\mathcal{V}$-optimal designs. The algorithms can be generalized to other variance-covariance structures as well. For example, Näther (1985) extends the algorithm of Wynn (1972) for general variance-covariance structures. Goos and Vandebroek (1997) use a rank 2 algorithm to compute $\mathcal{D}$-optimal designs for the simultaneous estimation of the mean and the variance function, as well as Goos, Tack and Vandebroek (2001), who describe an algorithm to generate $\mathcal{D}$-optimal designs for variance function estimation using sample variances. In this book, we will describe a number of exchange algorithms to cope with compound symmetric error structures.

2
Advanced Topics in Optimal Design

In many experimental situations, the assumptions of a homogeneous variance and uncorrelated observations are no longer satisfied. However, the design of experiments under these circumstances has only recently received attention. In this chapter, we give a concise overview of the work that has been done when the experimental observations do not have a constant variance and when the observations are correlated to each other. We also examine the design of experiments when the experimental units are heterogeneous. In that case, the experiment has to be blocked. Throughout this chapter, we show how the information matrix for each of the experiments can be written as a sum of outer products of vectors. This is extremely important for design construction algorithms because it allows a fast update of the information matrix, its determinant and its inverse.

2.1 Heterogeneous variance

As demonstrated by Taguchi (1989), reducing variability is an integral part of quality improvement. Therefore, industrial statisticians have become aware that they should not only be concerned with the expected value of the quality characteristic under investigation, but also with its variability. This led to a considerable interest in the simultaneous modeling of mean and variance in the field of engineering. The use of both the mean and the variance function has also gained popularity after Vining and Myers (1990) adapted the dual response approach to achieve the goals of Taguchi's philosophy. In this context, the assumption of homogeneous variance can no longer be held. Instead of ordinary least squares, the model parameters should then be estimated by generalized least squares (GLS):

$$\hat{\beta} = (\mathbf{X}'\mathbf{V}^{-1}\mathbf{X})^{-1}\mathbf{X}'\mathbf{V}^{-1}\mathbf{y}, \qquad (2.1)$$

where $\mathbf{V}$ is a diagonal matrix with the variances of the individual observations as the diagonal elements:

$$\begin{bmatrix} \sigma_1^2 & 0 & \cdots & 0 \\ 0 & \sigma_2^2 & \cdots & 0 \\ \vdots & \ddots & \ddots & \vdots \\ 0 & 0 & \cdots & \sigma_n^2 \end{bmatrix},$$

where σ_i^2 is the variance of the ith observation. The generalized least squares estimator is equivalent to the maximum likelihood estimator under normal errors. The variance-covariance matrix can be expressed as

$$\mathrm{Cov}(\hat{\beta}) = (\mathbf{X}'\mathbf{V}^{-1}\mathbf{X})^{-1}, \qquad (2.2)$$

and the information matrix on the unknown parameters is given by

$$\mathbf{M} = \mathbf{X}'\mathbf{V}^{-1}\mathbf{X}. \qquad (2.3)$$

When the variance components are unknown but estimated consistently, (2.3) is only asymptotically valid. From these expressions, it can be seen that the properties of the estimators depend on the design matrix $\mathbf{X}$ as well as on the variance-covariance matrix of the observations. As a result, optimal designs will also depend on the variance-covariance matrix and thus on the variances of the individual observations. Although the estimation of the variance function has received considerable attention in the literature, relatively little has been done on experimental design under heteroscedastic errors. Optimal designs for polynomial regression on one variable under specific heterogeneous variance structures have been obtained by Smith (1918), Schoenberg (1959), Karlin and Studden (1966) and Fedorov (1972). Mays and Easter (1997) derive optimal response surface designs in the presence of dispersion effects. Mays (1999) investigates how efficient central composite designs are when compared to designs optimally designed for a certain variance structure. Atkinson and Cook (1995), Vining and Schaub (1996), Goos and Vandebroek (1997) and Tack, Goos and Vandebroek (2002) propose $\mathcal{D}$-optimal designs for the simultaneous estimation of the mean and the variance function. In each of these papers, some prior knowledge about the magnitude of the variance is assumed. Also, it is implicitly assumed that residuals are used for the estimation of the variance function.

The exchange algorithms described in Section 1.9 can easily be adapted to develop $\mathcal{D}$-optimal designs in the presence of variance heterogeneity. In that case, the variance-covariance matrix $\mathbf{V}$ of the observations is diagonal and the information matrix can be expressed as a weighted sum of outer products:

$$\mathbf{M} = \sum_{i=1}^{n} \frac{\mathbf{f}(\mathbf{x}_i)\mathbf{f}'(\mathbf{x}_i)}{\sigma_i^2},$$

$$= \sum_{i=1}^{n} w_i \mathbf{f}(\mathbf{x}_i)\mathbf{f}'(\mathbf{x}_i), \tag{2.4}$$

where $w_i = \sigma_i^{-2}$. In order to update the information matrix after addition or deletion of an observation in design point $\mathbf{a}$, the weighted outer product $w_a \mathbf{f}(\mathbf{a})\mathbf{f}'(\mathbf{a})$ is added or subtracted. The modified $\mathcal{D}$-criterion value equals

$$|\mathbf{M} \pm w_a \mathbf{f}(\mathbf{a})\mathbf{f}'(\mathbf{a})| = |\mathbf{M}|\{1 \pm w_a \mathbf{f}'(\mathbf{a})\mathbf{M}^{-1}\mathbf{f}(\mathbf{a})\}. \tag{2.5}$$

Exchanging point $\mathbf{b}$ by point $\mathbf{a}$ modifies the $\mathcal{D}$-criterion value by a factor

$$\{1 + w_a \mathbf{f}'(\mathbf{a})\mathbf{M}^{-1}\mathbf{f}(\mathbf{a})\}\{1 - w_b \mathbf{f}'(\mathbf{b})\mathbf{M}^{-1}\mathbf{f}(\mathbf{b})\} + w_a w_b \{\mathbf{f}'(\mathbf{a})\mathbf{M}^{-1}\mathbf{f}(\mathbf{b})\}^2. \tag{2.6}$$

The inverse of the information matrix after addition or deletion of a design point $\mathbf{a}$ is

$$\{\mathbf{M} \pm \mathbf{f}(\mathbf{a})\mathbf{f}'(\mathbf{a})\}^{-1} = (\mathbf{X}'\mathbf{X})^{-1} \mp w_a \frac{\{\mathbf{M}^{-1}\mathbf{f}(\mathbf{a})\}\{\mathbf{M}^{-1}\mathbf{f}(\mathbf{a})\}'}{1 \pm w_a \mathbf{f}'(\mathbf{a})\mathbf{M}^{-1}\mathbf{f}(\mathbf{a})}. \tag{2.7}$$

Applications of these update formulae can be found in Vining and Schaub (1996), Goos and Vandebroek (1997) and Goos, Tack and Vandebroek (2001). They are also extensively used in the exchange algorithms developed in Chapters 4, 6, 7 and 8.

2.2   Correlated observations

Often, the experimental observations are correlated. For example, in repeated measurement studies, it is natural to assume that all observations carried out on one subject are correlated. In blocked experiments, observations belonging to the same block can be correlated as well. Similarly, when collecting spatial data, observations lying close to each other will not be statistically independent of each other. Spatial data occur in agriculture, geology and environmental sciences, among others. For model estimation, ordinary least squares is not the best choice in these cases and generalized least squares should be used. From (2.2) and (2.3), it can be seen that the optimal designs now depend on the correlation structure.

Martin, Jones and Eccleston (1998) and Martin, Eccleston and Jones (1998) investigate how two- and multi-level factorial designs should be used under different correlation structures like AR(1) and MA(1). Bischoff (1993) derives conditions under which an exact $\mathcal{D}$-optimal design for uncorrelated observations and homogeneous variance is also $\mathcal{D}$-optimal for correlated observations. Berger and Tan (1998) compute $\mathcal{D}$-optimal repeated measurement designs with first order autoregressive correlations and compare them to designs with equally spaced time points. Cross-over designs for treatment comparison when the observations are statistically dependent receive attention in Kunert (1991) and Donev (1998b). Eccleston and Chan (1998) use simulated annealing and tabu search to search for optimal row-column designs with correlated observations, whereas Donev (1998a) uses an exchange algorithm for the same purpose. The design of experiments in the presence of random block effects are treated by Chasalow (1992), Atkins (1994), Cheng (1995), Atkins and Cheng (1999) and Goos and Vandebroek (2001a), while Letsinger, Myers and Lentner (1996) and Goos and Vandebroek (2001c) consider split-plot experiments. Finally, Näther (1985) and Müller (1998) describe the design of random field experiments and take into account the correlation structure of the experiment.

The ideas of Fedorov (1972) can be used to construct optimal cross-over designs, even when the observations are no longer statistically independent. However, updating the determinant of the information matrix, that is the $\mathcal{D}$-criterion value, and the inverse information matrix becomes more cumbersome. This is due to the fact that, as soon as the observations are correlated, the information matrix can no longer be written as a weighted sum of the outer products of the polynomial expansions $\mathbf{f}(\mathbf{x})$ of the design points. Cholesky decomposition (see Theorem 14.5.11 of Harville (1997)) of the variance-covariance matrix $\mathbf{V}$ still allows us to write the information matrix as a sum of outer products, but the individual components of the resulting expression do no longer have a meaningful interpretation. This is shown in Chasalow (1992) and Donev (1998b). Consider for example the problem of designing a cross-over experiment with k observations on each of b subjects. This design problem is discussed in Donev (1998b). The information matrix of this type of experiment is of the form (2.3) with $\mathbf{V}$ a block diagonal matrix. Let us denote

$$
\mathbf{V} = \begin{bmatrix} \mathbf{V}_1 & \mathbf{0} & \cdots & \mathbf{0} \\ \mathbf{0} & \mathbf{V}_2 & \cdots & \mathbf{0} \\ \vdots & \ddots & \ddots & \vdots \\ \mathbf{0} & \mathbf{0} & \cdots & \mathbf{V}_b \end{bmatrix},
$$

where $\mathbf{V}_i$ is the variance-covariance matrix of all observations on the ith subject, and

$$
\mathbf{X} = \begin{bmatrix} \mathbf{X}_1' & \mathbf{X}_2' & \cdots & \mathbf{X}_b' \end{bmatrix}',
$$

where $\mathbf{X}_i$ is the part of the total design matrix $\mathbf{X}$ containing the treatments for the ith subject. Since

$$\mathbf{V}^{-1} = \begin{bmatrix} \mathbf{V}_1^{-1} & \mathbf{0} & \cdots & \mathbf{0} \\ \mathbf{0} & \mathbf{V}_2^{-1} & \cdots & \mathbf{0} \\ \vdots & \ddots & \ddots & \vdots \\ \mathbf{0} & \mathbf{0} & \cdots & \mathbf{V}_b^{-1} \end{bmatrix},$$

the information matrix can be written as

$$\mathbf{M} = \sum_{i=1}^{b} \mathbf{X}_i' \mathbf{V}_i^{-1} \mathbf{X}_i. \tag{2.8}$$

Let $\mathbf{L}_i$ be the lower triangular matrix resulting from the Cholesky decomposition of $\mathbf{V}_i^{-1}$ so that $\mathbf{V}_i^{-1} = \mathbf{L}_i \mathbf{L}_i'$. The information matrix then becomes

$$\begin{aligned} \mathbf{M} &= \sum_{i=1}^{b} \mathbf{X}_i' \mathbf{L}_i \mathbf{L}_i' \mathbf{X}_i, \\ &= \sum_{i=1}^{b} (\mathbf{L}_i' \mathbf{X}_i)' \mathbf{L}_i' \mathbf{X}_i, \\ &= \sum_{i=1}^{b} \mathbf{T}_i' \mathbf{T}_i, \\ &= \sum_{i=1}^{b} \sum_{j=1}^{k} \mathbf{t}_{ij} \mathbf{t}_{ij}', \end{aligned} \tag{2.9}$$

where $\mathbf{T}_i = \mathbf{L}_i' \mathbf{X}_i$ and $\mathbf{t}_{ij}'$ is the jth row of $\mathbf{T}_i$. As a consequence, the information matrix can still be written as a sum of outer products.

In the case of uncorrelated observations, updating the information matrix after addition or deletion of a design point was done by adding or subtracting the (weighted) outer products of the polynomial expansions of the design points (see Sections 1.8 and 2.1). As a result, updating the information matrix after exchanging a design point could be accomplished by addition and subtraction of an outer product. In the present case of correlated observations, a design change for the ith subject affects the partial design matrix $\mathbf{X}_i$ and hence each element of $\mathbf{T}_i$. As a result, k outer products $\mathbf{t}_{ij} \mathbf{t}_{ij}'$ have to be subtracted and k new outer products have to be added. It is clear that this is computationally much more prohibitive than in the uncorrelated case. In addition, updating the determinant and the inverse of the information matrix will be much more complicated as well.

Rather than considering exchanges of design points, Donev (1998b) tries to improve the initial starting design by exchanging blocks of observations. This adds flexibility to the design construction algorithm without increasing the computational burden because each design change for the ith subject, no matter whether one or k treatments are replaced, entails $2k$ operations to update the information matrix. The drawback of this approach is that it requires a complete enumeration of all possible matrices $\mathbf{X}_i$. The number of different $\mathbf{X}_i$ is equal to h^k, where h is the number of possible factor level combinations, or equivalently the number of candidate design points, and k is the number of observations performed on each of the subjects. Of course, the number of different $\mathbf{X}_i$ rapidly increases with increasing number of different factor level combinations and with increasing block size. For quadratic regression on one explanatory variable, it is not uncommon to use three levels for the factor under investigation: -1, 0 and +1. With blocks of size two, $3^2 = 9$ different $\mathbf{X}_i$ have to be considered for general $\mathbf{V}_i$. Note that, unlike in the case of uncorrelated observations, the sequence of the experimental runs has an impact on the design efficiency if the observations are correlated. When the effect of three factors is investigated in an experiment and a quadratic model has to be estimated, the 27 points of the 3^3 factorial design are often used as candidate design points. When the block size equals 4, the total number of different blocks amounts to $27^4 = 531441$. When the sequence of the observations within one subject or block does not matter, for example when the error structure is compound symmetric (see Chapter 3), all $\mathbf{X}_i$ that can be obtained from other $\mathbf{X}_i$ by a permutation of its rows are equivalent. In that case, only

$$\binom{h + k - 1}{k}$$

matrices $\mathbf{X}_i$ need to be enumerated. The number of $\mathbf{X}_i$ needed is then equal to 6 for a problem with $h = 3$ and $k = 2$ and to 27405 when $h = 27$ and $k = 4$. The enumeration when the error structure is compound symmetric comes down to generating the points of a simplex lattice. An algorithm for this purpose was developed by Chasalow and Brand (1995). This algorithm was used in Chasalow (1992) to generate all possible $\mathbf{X}_i$ and to compose all possible designs for an experiment with block size two and one explanatory variable at three levels. It is clear that, in contrast with Donev (1998b), Chasalow (1992) does not use a design construction algorithm but he enumerates all possible designs. Suppose that b subjects are available and that there are n_d possible blocks, then

$$\binom{n_d + b - 1}{b}$$

different designs are evaluated.

It is clear that enumerating all possible $\mathbf{X}_i$ is impractical and time consuming. This is especially so when the number of factor level combinations under consideration is large. For response surface models with more than one experimental variable and with a couple of quadratic terms, this large number is inevitable. In addition, the block size is often heterogeneous so that all possible $\mathbf{X}_i$ for each block size have to be enumerated. Therefore, for the design problems considered in this book, we have developed point exchange algorithms instead of block exchange algorithms. The point exchange algorithms developed in this text exploit the compound symmetric error structure, which is common to all bi-randomization or two-stratum experiments. One of the bi-randomization experiments considered in this book is an experiment with random block effects. In the next section, we will describe the statistical models corresponding to blocked experiments and pay attention to the differences between random and fixed block effects.

2.3   Blocking experiments

The available experimental units are in many cases heterogeneous because it is often impossible to perform all the experimental runs under homogeneous conditions. In order to attain homogeneous conditions, it may be necessary to perform all runs in a single time period (such as one day), to use a single batch of raw material, to use only one operator, etc. When this is impossible, experiments are blocked so that the experimental units within the blocks are more homogeneous than experimental units from different blocks. For example, in the chemical industry, the yield of a process depends not only on the settings of the experimental variables but also on the batch of raw material used. Such an extraneous source of variation is referred to as a block effect. In a blocked experiment, part of the variance of the observations can be attributed to the blocks, decreasing the remaining variance. Therefore, the effects of the experimental variables can be estimated more precisely. As a result, blocking an experiment allows a more powerful analysis. Examples of blocked experiments with qualitative factors were already given in Section 1.6.

In this section, we will describe two commonly used linear models with additive block effects. We assume that there is no interaction between the experimental variables and the blocks.

2.3.1   The fixed and the random block effects model

The statistical model corresponding to a blocked experiment can be written as

$$y_{ij} = \mathbf{f}'(\mathbf{x}_{ij})\beta + \gamma_i + \varepsilon_{ij} \tag{2.10}$$

where y_{ij} is the response of the jth observation in the ith block, $\mathbf{f}(\mathbf{x}_{ij})$ is the polynomial expansion of the factor level combination $\mathbf{x}_{ij}$, β is the

$p \times 1$ vector containing the p effects of the experimental variables, γ_i is the ith block effect, and ε_{ij} is the random error. Note that β in many cases contains an intercept β_0. The number of blocks in the experiment is denoted by b and the number of observations within the ith block is equal to k_i. In matrix notation, the model becomes

$$\mathbf{y} = \mathbf{X}\beta + \mathbf{Z}\gamma + \varepsilon, \tag{2.11}$$

where $\mathbf{y}$ is a vector of n observations on the response of interest, the vector β contains the p unknown fixed parameters, the vector $\gamma = [\gamma_1, \gamma_2, \dots, \gamma_b]'$ contains the b block effects and ε is a random error vector. The matrices $\mathbf{X}$ and $\mathbf{Z}$ are known and have dimension $n \times p$ and $n \times b$ respectively. $\mathbf{X}$ contains the polynomial expansions of the m factor levels at the n experimental runs. If the runs of the experiment are grouped per block, then $\mathbf{Z}$ is of the form

$$\mathbf{Z} = \mathrm{diag}[\mathbf{1}_{k_1}, \mathbf{1}_{k_2}, \dots, \mathbf{1}_{k_b}], \tag{2.12}$$

where $\mathbf{1}_k$ is a $k \times 1$ vector of ones.

In the fixed block effects model, no distribution is assumed for the block effects. In that case, the only distributional assumption is concerned with the random errors:

$$\mathrm{E}(\varepsilon) = \mathbf{0}_n \text{ and } \mathrm{Cov}(\varepsilon) = \sigma_\varepsilon^2 \mathbf{I}_n. \tag{2.13}$$

In the random block effects model, the block effects are assumed to be randomly drawn from a distribution. The distributional assumptions are given by

$$\mathrm{E}(\varepsilon) = \mathbf{0}_n \text{ and } \mathrm{Cov}(\varepsilon) = \sigma_\varepsilon^2 \mathbf{I}_n, \tag{2.14}$$

$$\mathrm{E}(\gamma) = \mathbf{0}_b \text{ and } \mathrm{Cov}(\gamma) = \sigma_\gamma^2 \mathbf{I}_b, \tag{2.15}$$

$$\text{and } \mathrm{Cov}(\gamma, \varepsilon) = \mathbf{0}_{b \times n}. \tag{2.16}$$

2.3.2 Model choice

In principle, the choice between fixed and random blocks should be guided by how the blocks were selected. If the blocks are chosen by design, using fixed block effects is appropriate. If, however, the blocks constitute a random sample from a population of blocks, then using random block effects is appropriate. In practice, this principle is seldom used. Partly, this is due to the fact that the distinction between random and fixed blocks is not always very clear. Ganju (2000) also points out that many experimenters prefer the random block effects model independently of the fashion in which the blocks were selected. This is because the analysis of a model with random blocks results in smaller variances of the treatment contrasts when a balanced incomplete block design is used. Gilmour and Trinca (2000) would also treat the blocks as random if the block labels are randomly allocated to the blocks. Since this random allocation is standard practice, such a suggestion is almost a recommendation to treat the blocks always as random.

The analysis of an experiment with fixed block effects is called intra-block analysis and is a special case of fixed effects analysis. The analysis of an experiment with random block effects is referred to as combined intra- and interblock analysis or random blocks analysis, which is a special case of mixed effects analysis. The latter analysis is more efficient because it recovers the so-called interblock information from the experiment. This is illustrated by Gilmour and Trinca (2000), among others. Ganju (2000) recommends, under some conditions, the fixed effects analysis for comparing treatments, that is qualitative factors, in order to avoid the risk of ending up with a negative estimate for σ_γ^2. This is because there is no consensus on how to deal adequately with the negative estimate. One solution is to ignore the block effect. Another is to treat the blocks as fixed. When the experimental factors are continuous, Ganju (2000) recommends the random blocks analysis because there is no unique solution to the variance of a predicted response. Khuri (1992) as well as Ganju (2000) and Gilmour and Trinca (2000) point out that the fixed effects analysis is equivalent to the random blocks analysis when $\sigma_\gamma^2 \to \infty$. As a result, the larger the variance between blocks σ_γ^2, the smaller will be the benefit of using the random blocks analysis.

The random block effects model is treated extensively in Chapter 4. In the sequel of this section, we will concentrate on the fixed block effects model, that is the model given by (2.11) and (2.13). Firstly, we will describe the analysis. Next, we will introduce the concept of orthogonal blocking and show that orthogonal blocking is an optimal strategy for assigning the points or treatments of a given design matrix $\mathbf{X}$ to the blocks.

2.3.3   Intra-block analysis

When model (2.11) contains an intercept, the design matrix $\mathbf{X}$ has a column of ones and can be written as $[\ \mathbf{1}_n\ \tilde{\mathbf{X}}\ ]$. Since summing the columns of $\mathbf{Z}$ also yields a column of ones, the rank of the matrix $[\ \mathbf{X}\ \mathbf{Z}\ ]$ is then less than $p + b$, the number of parameters in the model. As a result, the intercept and the block effects in the fixed block effects model cannot be estimated independently of each other. Therefore, the model (2.11) is written as

$$
\begin{aligned}
\mathbf{y} &= \tilde{\mathbf{X}}\tilde{\boldsymbol{\beta}} + \mathbf{Z}\boldsymbol{\tau} + \boldsymbol{\varepsilon}, \\
&= \mathbf{W}\boldsymbol{\theta} + \boldsymbol{\varepsilon},
\end{aligned}
\tag{2.17}
$$

where $\tilde{\boldsymbol{\beta}}$ is the $(p-1)$-dimensional vector containing all elements of $\boldsymbol{\beta}$ apart from the intercept β_0, $\boldsymbol{\tau} = \beta_0\mathbf{1}_b + \boldsymbol{\gamma}$, $\mathbf{W} = [\ \tilde{\mathbf{X}}\ \mathbf{Z}\ ]$, and $\boldsymbol{\theta}' = [\ \tilde{\boldsymbol{\beta}}'\ \boldsymbol{\tau}'\ ]$. The ordinary least squares estimate of $\boldsymbol{\theta}$ is given by

$$\hat{\boldsymbol{\theta}} = \begin{bmatrix} \hat{\tilde{\boldsymbol{\beta}}} \\ \hat{\boldsymbol{\tau}} \end{bmatrix} = (\mathbf{W}\mathbf{W})^{-1}\mathbf{W}\mathbf{y}$$

$$= \begin{bmatrix} \tilde{\mathbf{X}}'\tilde{\mathbf{X}} & \tilde{\mathbf{X}}'\mathbf{Z} \\ \mathbf{Z}'\tilde{\mathbf{X}} & \mathbf{Z}'\mathbf{Z} \end{bmatrix}^{-1} \begin{bmatrix} \tilde{\mathbf{X}}' \\ \mathbf{Z}' \end{bmatrix} \mathbf{y},$$

(2.18)

where $\hat{\boldsymbol{\beta}} = [\ \hat{\beta}_0\ \hat{\boldsymbol{\beta}}'\ ]'$. Applying Theorem 8.5.11 of Harville (1997) to this equation and using the facts that

$$\mathbf{Z}'\mathbf{Z} = \mathrm{diag}[k_1, k_2, \ldots, k_b],$$ (2.19)

so that

$$(\mathbf{Z}'\mathbf{Z})^{-1} = \mathrm{diag}[k_1^{-1}, k_2^{-1}, \ldots, k_b^{-1}],$$ (2.20)

$$\mathbf{Z}'\mathbf{y} = \begin{bmatrix} \sum_{j=1}^{k_1} y_{1j} \\ \sum_{j=1}^{k_2} y_{2j} \\ \vdots \\ \sum_{j=1}^{k_b} y_{bj} \end{bmatrix},$$ (2.21)

and

$$\mathbf{Z}'\tilde{\mathbf{X}} = \begin{bmatrix} \mathbf{1}'_{k_1}\tilde{\mathbf{X}}_1 \\ \mathbf{1}'_{k_2}\tilde{\mathbf{X}}_2 \\ \vdots \\ \mathbf{1}'_{k_b}\tilde{\mathbf{X}}_b \end{bmatrix},$$ (2.22)

gives us an expression for the estimators of the block effects:

$$\hat{\boldsymbol{\tau}} = (\mathbf{Z}'\mathbf{Z})^{-1}\mathbf{Z}'\mathbf{y} - (\mathbf{Z}'\mathbf{Z})^{-1}\mathbf{Z}'\tilde{\mathbf{X}}\hat{\tilde{\boldsymbol{\beta}}},$$

$$= \begin{bmatrix} \bar{y}_1 \\ \bar{y}_2 \\ \vdots \\ \bar{y}_b \end{bmatrix} - \begin{bmatrix} \mathbf{1}'_{k_1}\tilde{\mathbf{X}}_1\hat{\tilde{\boldsymbol{\beta}}}/k_1 \\ \mathbf{1}'_{k_2}\tilde{\mathbf{X}}_2\hat{\tilde{\boldsymbol{\beta}}}/k_2 \\ \vdots \\ \mathbf{1}'_{k_b}\tilde{\mathbf{X}}_b\hat{\tilde{\boldsymbol{\beta}}}/k_b \end{bmatrix},$$ (2.23)

where $\bar{y}_i$ is the average response of the observations in the ith block. A prediction for the mean response for a factor level combination $\mathbf{x}$ is given by

$$\hat{y} = \hat{\tau}_i + \tilde{\mathbf{f}}'(\mathbf{x})\hat{\tilde{\boldsymbol{\beta}}},$$ (2.24)

where $\mathbf{f}'(\mathbf{x}) = [\ 1\ \tilde{\mathbf{f}}'(\mathbf{x})\ ]$. It should be emphasized that $\hat{\tau}_i$ is unambiguously determined and that $\hat{y}$ is therefore unique. This is in contrast with

Khuri's (1994) suggestion to use

$$\hat{y} = \hat{\beta}_0 + \tilde{\mathbf{f}}'(\mathbf{x})\hat{\tilde{\boldsymbol{\beta}}} \qquad (2.25)$$

as a prediction. This approach is inconsistent with the assumption of fixed blocks because assuming fixed blocks is equivalent to assuming that the response in each block is different and unpredictable. The inferences made then only apply to the blocks actually used in the experiment and cannot be extrapolated without additional assumptions. In addition, Khuri's prediction depends on the estimate for the intercept. As already pointed out, the intercept cannot be estimated independently of the block effects. Therefore, a constraint has to be imposed on the block effects $\boldsymbol{\gamma}$ or $\boldsymbol{\tau}$ if an estimate for the intercept β_0 is needed. The choice of the constraint has an impact on the estimate of β_0, so that, in Khuri's approach, there is no unique prediction in the fixed block effects model. If the block effects are constrained to sum to zero, that is

$$\sum_{i=1}^{b} \gamma_i = \mathbf{1}_b' \boldsymbol{\gamma} = 0, \qquad (2.26)$$

then

$$\hat{\beta}_0 = \frac{1}{b} \sum_{i=1}^{b} \hat{\tau}_i = \frac{1}{b} \sum_{i=1}^{b} \bar{y}_i - \frac{1}{b} \sum_{i=1}^{b} \mathbf{1}_{k_i}' \tilde{\mathbf{X}}_i \hat{\tilde{\boldsymbol{\beta}}} / k_i. \qquad (2.27)$$

This approach is used in Khuri (1994). Another possibility is to constrain the block effects so that the weighted sum of the block effects is zero:

$$\sum_{i=1}^{b} k_i \gamma_i = \mathbf{k}' \boldsymbol{\gamma} = 0, \qquad (2.28)$$

where $\mathbf{k} = [\, k_1 \ k_2 \ \ldots \ k_b \,]'$. In that case, an expression for the intercept is given by

$$\hat{\beta}_0 = \frac{1}{n} \sum_{i=1}^{b} k_i \hat{\tau}_i = \bar{y} - \frac{1}{n} \mathbf{1}_n' \tilde{\mathbf{X}} \hat{\tilde{\boldsymbol{\beta}}}. \qquad (2.29)$$

This approach is used in Trinca and Gilmour (2000) and always leads to the estimate for the intercept that would be obtained by ignoring the blocks. When the block sizes are homogeneous, that is $k = k_1 = k_2 = \cdots = k_b$, (2.27) and (2.29) are equivalent. However, they produce different estimates when the block sizes are heterogeneous. What constraint should be used depends on the experimental situation. If the intercept is interpreted as the overall mean of the experimental observations, then the second constraint is correct. If, however, the overall population mean is meant, the choice of the constraint should depend on the population block sizes. If the population block sizes are assumed to be proportional to the block sizes of the experiment, the second constraint is correct. If the population block sizes are assumed to be homogeneous, unlike the block sizes of the experiment, the first constraint is correct.

Of course, not only the estimate of the intercept depends on the constraint imposed, but also the prediction variance. This discussion is used by Ganju (2000) as an argument in favor of random blocks analysis. This argument becomes, however, invalid when the interest is in predicting functions of the response, for example the difference between two responses or the location of a stationary point. This is because the prediction then no longer depends on the intercept, nor on the block effects.

The variance-covariance matrix of the ordinary least squares estimator $\hat{\theta}$ is

$$\text{var}(\hat{\theta}) = \sigma_\varepsilon^2 \begin{bmatrix} \tilde{X}'\tilde{X} & \tilde{X}'Z \\ Z'\tilde{X} & Z'Z \end{bmatrix}^{-1}, \tag{2.30}$$

and the information matrix on the unknown parameter θ equals

$$\sigma_\varepsilon^{-2} \begin{bmatrix} \tilde{X}'\tilde{X} & \tilde{X}'Z \\ Z'\tilde{X} & Z'Z \end{bmatrix}. \tag{2.31}$$

Using Harville's (1997) Theorem 8.5.11 on the inverse of a partitioned matrix, we find that

$$\text{var}(\hat{\tilde{\beta}}) = \sigma_\varepsilon^2 \{\tilde{X}'\tilde{X} - \tilde{X}'Z(Z'Z)^{-1}Z'\tilde{X}\}^{-1}. \tag{2.32}$$

When we denote the portion of $\tilde{X}$ corresponding to the ith block by $\tilde{X}_i$, we have that

$$\begin{aligned} \tilde{X}'Z &= [\,\tilde{X}_1' \quad \tilde{X}_2' \quad \ldots \quad \tilde{X}_b'\,]\,\text{diag}[1_{k_1}, 1_{k_2}, \ldots, 1_{k_b}\,], \\ &= [\,\tilde{X}_1' 1_{k_1} \quad \tilde{X}_2' 1_{k_2} \quad \ldots \quad \tilde{X}_b' 1_{k_b}\,]. \end{aligned} \tag{2.33}$$

Hence

$$\tilde{X}'Z(Z'Z)^{-1} = [\,\tilde{X}_1' 1_{k_1}/k_1 \quad \tilde{X}_2' 1_{k_2}/k_2 \quad \ldots \quad \tilde{X}_b' 1_{k_b}/k_b\,], \tag{2.34}$$

and

$$\tilde{X}'Z(Z'Z)^{-1}Z'\tilde{X} = \sum_{i=1}^{b} \frac{1}{k_i}(\tilde{X}_i' 1_{k_i})(\tilde{X}_i' 1_{k_i})'. \tag{2.35}$$

As a result,

$$\text{var}(\hat{\tilde{\beta}}) = \sigma_\varepsilon^2 \{\tilde{X}'\tilde{X} - \sum_{i=1}^{b} \frac{1}{k_i}(\tilde{X}_i' 1_{k_i})(\tilde{X}_i' 1_{k_i})'\}^{-1}. \tag{2.36}$$

The information matrix on the unknown $\tilde{\beta}$ is given by

$$M = \sigma_\varepsilon^{-2} \{\tilde{X}'\tilde{X} - \sum_{i=1}^{b} \frac{1}{k_i}(\tilde{X}_i' 1_{k_i})(\tilde{X}_i' 1_{k_i})'\}. \tag{2.37}$$

2.3.4   Orthogonal blocking

In general, the polynomial effects cannot be estimated independently of the block effects. It is, however, desirable that the estimates for the polynomial effects $\hat{\beta}$ are not influenced by the estimation of the block effects γ. Any design achieving this objective is said to be orthogonally blocked. For this purpose, the columns of

$$\bar{Z} = (I_n - n^{-1}1_n1_n')Z$$

should be orthogonal to the columns of X that do not correspond to the intercept, that is $\tilde{X}$. This can be seen by rewriting the fixed block effects model (2.11) as

$$y = \mu 1_n + \tilde{X}\tilde{\beta} + \bar{Z}\gamma$$

where $\mu = \beta_0 + n^{-1}1_n'Z\gamma$. The matrices $\bar{Z}$ and $\tilde{X}$ are orthogonal when $\bar{Z}'\tilde{X} = 0_{b\times(p-1)}$. Now,

$$
\begin{aligned}
\bar{Z}'\tilde{X} &= Z'(I_n - n^{-1}1_n1_n')'\tilde{X}, \\
&= Z'(\tilde{X} - n^{-1}1_n1_n'\tilde{X}), \\
&= Z'\tilde{X} - n^{-1}Z'1_n1_n'\tilde{X}, \\
&= Z'\tilde{X} - n^{-1}k1_n'\tilde{X}, \\
&= \begin{bmatrix} 1_{k_1}'\tilde{X}_1 \\ 1_{k_2}'\tilde{X}_2 \\ \vdots \\ 1_{k_b}'\tilde{X}_b \end{bmatrix} - \frac{1}{n}\begin{bmatrix} k_1 \\ k_2 \\ \vdots \\ k_b \end{bmatrix}1_n'\tilde{X}.
\end{aligned}
\tag{2.38}
$$

As a result, the condition for orthogonality can also be written as

$$\tilde{X}_i'1_{k_i} = \frac{k_i}{n}\tilde{X}'1_n, \qquad (i = 1, 2, \ldots, b). \tag{2.39}$$

Since $1_{k_i}'1_{k_i} = k_i1_n'1_n/n$, we have that

$$\begin{bmatrix} 1_{k_i}'1_{k_i} \\ \tilde{X}_i'1_{k_i} \end{bmatrix} = \frac{k_i}{n}\begin{bmatrix} 1_n'1_n \\ \tilde{X}'1_n \end{bmatrix}, \qquad (i = 1, 2, \ldots, b), \tag{2.40}$$

or

$$\frac{1}{k_i}X_i'1_{k_i} = \frac{1}{n}X'1_n, \qquad (i = 1, 2, \ldots, b). \tag{2.41}$$

In words, the average row of all X_i is the same for all blocks and it is equal to the average row of the total design matrix X. From (2.38), it can be seen that $Z'\tilde{X} = n^{-1}k1_n'\tilde{X}$ when the orthogonality condition is satisfied.

Using Harville's (1997) Theorem 8.5.11 on the inverse of a partitioned matrix, the intra-block estimator (2.18) can be rewritten as

$$\begin{bmatrix} \hat{\tilde{\beta}} \\ \hat{\tau} \end{bmatrix} = \begin{bmatrix} \mathbf{Q}_2 & -\mathbf{Q}_2\tilde{\mathbf{X}}'\mathbf{Z}(\mathbf{Z}'\mathbf{Z})^{-1} \\ -(\mathbf{Z}'\mathbf{Z})^{-1}\mathbf{Z}'\tilde{\mathbf{X}}\mathbf{Q}_2 & (\mathbf{Z}'\mathbf{Z})^{-1} + (\mathbf{Z}'\mathbf{Z})^{-1}\mathbf{Z}'\tilde{\mathbf{X}}\mathbf{Q}_1 \end{bmatrix} \begin{bmatrix} \tilde{\mathbf{X}}'\mathbf{y} \\ \mathbf{Z}'\mathbf{y} \end{bmatrix},$$

$$= \begin{bmatrix} \mathbf{Q}_2\tilde{\mathbf{X}}'\mathbf{y} - \mathbf{Q}_2\tilde{\mathbf{X}}'\mathbf{Z}(\mathbf{Z}'\mathbf{Z})^{-1}\mathbf{Z}'\mathbf{y} \\ -(\mathbf{Z}'\mathbf{Z})^{-1}\mathbf{Z}'\tilde{\mathbf{X}}\mathbf{Q}_2\tilde{\mathbf{X}}'\mathbf{y} + (\mathbf{Z}'\mathbf{Z})^{-1}\mathbf{Z}'\mathbf{y} + (\mathbf{Z}'\mathbf{Z})^{-1}\mathbf{Z}'\tilde{\mathbf{X}}\mathbf{Q}_1\mathbf{Z}'\mathbf{y} \end{bmatrix},$$

where

$$\mathbf{Q}_1 = \mathbf{Q}_2\tilde{\mathbf{X}}'\mathbf{Z}(\mathbf{Z}'\mathbf{Z})^{-1},$$

and

$$\mathbf{Q}_2 = \{\tilde{\mathbf{X}}'\tilde{\mathbf{X}} - \tilde{\mathbf{X}}'\mathbf{Z}(\mathbf{Z}'\mathbf{Z})^{-1}\mathbf{Z}'\mathbf{X}\}^{-1}.$$

Substituting $\mathbf{Z}'\tilde{\mathbf{X}} = n^{-1}\mathbf{k}\mathbf{1}_n'\tilde{\mathbf{X}}$ and $\mathbf{k}'(\mathbf{Z}'\mathbf{Z})^{-1} = \mathbf{1}_b'$ yields the following estimator for β:

$$\begin{aligned} \hat{\tilde{\beta}}_{\text{orth.}} &= \{\tilde{\mathbf{X}}'\tilde{\mathbf{X}} - n^{-2}\tilde{\mathbf{X}}'\mathbf{1}_n\mathbf{k}'(\mathbf{Z}'\mathbf{Z})^{-1}\mathbf{k}\mathbf{1}_n'\tilde{\mathbf{X}}\}^{-1} \\ &\quad \times \{\tilde{\mathbf{X}}'\mathbf{y} - n^{-1}\tilde{\mathbf{X}}'\mathbf{1}_n\mathbf{k}'(\mathbf{Z}'\mathbf{Z})^{-1}\mathbf{Z}'\mathbf{y}\}, \\ &= \{\tilde{\mathbf{X}}'\tilde{\mathbf{X}} - n^{-2}\tilde{\mathbf{X}}'\mathbf{1}_n\mathbf{1}_b'\mathbf{k}\mathbf{1}_n'\tilde{\mathbf{X}}\}^{-1}(\tilde{\mathbf{X}}'\mathbf{y} - n^{-1}\tilde{\mathbf{X}}'\mathbf{1}_n\mathbf{1}_b'\mathbf{Z}'\mathbf{y}), \\ &= (\tilde{\mathbf{X}}'\tilde{\mathbf{X}} - n^{-1}\tilde{\mathbf{X}}'\mathbf{1}_n\mathbf{1}_n'\tilde{\mathbf{X}})^{-1}(\tilde{\mathbf{X}}'\mathbf{y} - n^{-1}\tilde{\mathbf{X}}'\mathbf{1}_n\mathbf{1}_n'\mathbf{y}), \\ &= \{\tilde{\mathbf{X}}'(\mathbf{I}_n - n^{-1}\mathbf{1}_n\mathbf{1}_n')\tilde{\mathbf{X}}\}^{-1}\tilde{\mathbf{X}}'(\mathbf{I}_n - n^{-1}\mathbf{1}_n\mathbf{1}_n')\mathbf{y}. \end{aligned}$$

This estimator is identical to the estimator that would be obtained by ignoring the blocks. This was proven by Khuri (1992) and can be seen by replacing $\mathbf{Z}$ by $\mathbf{1}_n$ in (2.18) and applying an analogous reasoning as the one above. As a result, the estimator for $\tilde{\beta}$ is the same as that obtained from model (1.3), that is the model without block effects. From (2.36), we also have that

$$\text{var}(\hat{\tilde{\beta}}_{\text{orth.}}) = \sigma_\varepsilon^2\{\tilde{\mathbf{X}}'\tilde{\mathbf{X}} - n^{-1}(\tilde{\mathbf{X}}'\mathbf{1}_n)(\tilde{\mathbf{X}}'\mathbf{1}_n)'\}^{-1}. \tag{2.42}$$

This variance-covariance matrix is identical to (1.29), which is the variance-covariance matrix obtained from the model without block effects. Therefore, it is said that, in an orthogonally blocked design, no information on $\tilde{\beta}$ is lost due to blocking. This does however not mean that ignoring the block effects has no consequences for the data analysis when an orthogonally blocked experiment is used. This is because the estimate for σ_ε^2 is substantially larger when the blocks in the experiment are ignored. This leads to larger standard errors for the parameter estimates and makes it more difficult to detect significant effects.

2.3.5   Illustrations

Consider a central composite design (CCD) for 3 experimental variables with 6 replications of the center point and $\alpha = 1.6330$. The factorial por-

Table 2.1: Orthogonally blocked central composite design.

	x_1	x_2	x_3	x_1x_2	x_1x_3	x_2x_3	x_1^2	x_2^2	x_3^2
1	-1	-1	-1	1	1	1	1	1	1
2	1	1	-1	1	-1	-1	1	1	1
3	1	-1	1	-1	1	-1	1	1	1
4	-1	1	1	-1	-1	1	1	1	1
5	0	0	0	0	0	0	0	0	0
6	0	0	0	0	0	0	0	0	0
avg.	0	0	0	0	0	0	2/3	2/3	2/3
1	1	-1	-1	-1	-1	1	1	1	1
2	-1	1	-1	-1	1	-1	1	1	1
3	-1	-1	1	1	-1	-1	1	1	1
4	1	1	1	1	1	1	1	1	1
5	0	0	0	0	0	0	0	0	0
6	0	0	0	0	0	0	0	0	0
avg.	0	0	0	0	0	0	2/3	2/3	2/3
1	-1.63	0	0	0	0	0	2.67	0	0
2	1.63	0	0	0	0	0	2.67	0	0
3	0	-1.63	0	0	0	0	0	2.67	0
4	0	1.63	0	0	0	0	0	2.67	0
5	0	0	-1.63	0	0	0	0	0	2.67
6	0	0	1.63	0	0	0	0	0	2.67
7	0	0	0	0	0	0	0	0	0
8	0	0	0	0	0	0	0	0	0
avg.	0	0	0	0	0	0	2/3	2/3	2/3

tion of this design consists of $2^3 = 8$ points and the axial portion comprises $2 \times 3 = 6$ points, so that the total number of observations amounts to 20. This design can be blocked orthogonally when two blocks of six observations and one block of eight observations are available, that is when $b = 3$, $k_1 = k_2 = 6$ and $k_3 = 8$. The axial points plus two center runs are assigned to the largest block, while the two remaining blocks consist of a half fraction of the 2^3 factorial design plus two center runs. The resulting design is displayed in Table 2.1, along with the average rows of the individual blocks. Since the average row is the same for all blocks, the design is orthogonally blocked.

Another example of an orthogonally blocked central composite design is given in Table 2.2. The data are taken from a small reactor experiment described in Box and Draper (1987) and revisited in Khuri (1994). In the

Table 2.2: Orthogonally blocked central composite design with four blocks of size six.

Block	x_1	x_2	x_3	y
1	-1	-1	1	40.0
	1	-1	-1	18.6
	-1	1	-1	53.8
	1	1	1	64.2
	0	0	0	53.5
	0	0	0	52.7
2	-1	-1	-1	39.5
	1	-1	1	59.7
	-1	1	1	42.2
	1	1	-1	33.6
	0	0	0	54.1
	0	0	0	51.0
3	$-\alpha$	0	0	43.0
	α	0	0	43.9
	0	$-\alpha$	0	47.0
	0	α	0	62.8
	0	0	$-\alpha$	25.6
	0	0	α	49.7
4	$-\alpha$	0	0	39.2
	α	0	0	46.3
	0	$-\alpha$	0	44.9
	0	α	0	58.1
	0	0	$-\alpha$	27.0
	0	0	α	50.7

$\alpha = \sqrt{2}$

experiment, the effect of 3 factors (flow rate, concentration of a catalyst, and temperature) on the concentration of a product was investigated. The 24 runs of the experiment were performed sequentially in four blocks of size six. A full quadratic model was fitted to the data. The response of the jth observation in the ith block is then given by

$$y_{ij} = \beta_0 + \sum_{i=1}^{3} \beta_i x_i + \sum_{i=1}^{3} \sum_{j=i+1}^{3} \beta_{ij} x_i x_j + \sum_{i=1}^{3} \beta_{ii} x_i^2 + \gamma_i + \varepsilon_{ij}.$$

In order to perform an intra-block analysis on the data, the four blocks in the experiment are modelled as fixed blocks. The following SAS commands can be used to estimate the fixed block effects model:

```
data reactor;
infile 'reactor.dat';
input block x1 x2 x3 y;
x1x2=x1*x2;
```

Table 2.3: Coefficient estimates, standard errors, t-statistics and p-values for the small reactor experiment obtained by using the intra-block analyis.

Term	Estimate	St. error	t-value	p-value
Block 1	53.0500	1.0867	48.82	< .0001
Block 2	52.6000	1.0867	48.40	< .0001
Block 3	51.2499	1.0867	47.16	< .0001
Block 4	50.2832	1.0867	46.27	< .0001
x_1	0.7446	0.4706	1.58	0.1419
x_2	4.8133	0.4706	10.23	< .0001
x_3	8.0125	0.4706	17.03	< .0001
$x_1 x_2$	0.3750	0.6655	0.56	0.5844
$x_1 x_3$	10.3500	0.6655	15.55	< .0001
$x_2 x_3$	-2.8250	0.6655	-4.24	0.0014
x_1^2	-3.8334	0.5434	-7.05	< .0001
x_2^2	1.2167	0.5434	2.24	0.0468
x_3^2	-6.2584	0.5434	-11.52	< .0001

```
x1x3=x1*x3;
x2x3=x2*x3;
x11=x1*x1;
x22=x2*x2;
x33=x3*x3;
proc glm data = reactor;
class block;
model y = block x1 x2 x3 x1x2 x1x3 x2x3 x11 x22 x33
                              / solution noint;
run;
```

Note that the option NOINT was specified because the intercept and the four block effects cannot be estimated independent of each other. The estimated coefficients, as well as the estimated standard errors, the t-statistics and the p-values produced by SAS are displayed in Table 2.3.

Since the design is orthogonal, these parameter estimates can also be computed by ignoring the blocks, that is by using the following commands:

```
proc glm data = reactor;
model y = x1 x2 x3 x1x2 x1x3 x2x3 x11 x22 x33 / solution;
run;
```

The estimates, standard errors and t-statistics and p-values obtained by ignoring the block effects are displayed in Table 2.4. It can be verified that the estimates of the factor effects are identical to those obtained by using

Table 2.4: Coefficient estimates, standard errors, t-statistics and p-values for the small reactor experiment obtained by ignoring the blocks.

Term	Estimate	St. error	t-value	p-value
Intercept	51.7957	1.0045	51.56	$< .0001$
x_1	0.7446	0.5502	1.35	0.1974
x_2	4.8133	0.5502	8.75	$< .0001$
x_3	8.0125	0.5502	14.56	$< .0001$
$x_1 x_2$	0.3750	0.7781	0.48	0.6373
$x_1 x_3$	10.3500	0.7781	13.30	$< .0001$
$x_2 x_3$	-2.8250	0.7781	-3.63	0.0027
x_1^2	-3.8334	0.6353	-6.03	$< .0001$
x_2^2	1.2167	0.6353	1.92	0.0761
x_3^2	-6.2584	0.5434	-9.85	$< .0001$

the intra-block analysis. However, the standard errors and t-statistics obtained by ignoring the blocks are substantially larger than those produced by the intra-block analysis. This is because the residual variance is 3.5431 when the intra-block analysis is used and 4.8434 when the block effects are ignored. As a result, a larger part of the variance in the data remains unexplained when the block effects are omitted from the model. This makes it more difficult to detect significant effects. It can be verified from Tables 2.3 and 2.4 that the coefficient of x_2^2 is significantly different from zero at the 5% level when the intra-block analysis is used but not when the block effects are ignored. This example clearly shows that the blocks have to be taken into account when analyzing a blocked experiment, even when the experiment is orthogonally blocked. Finally, it can also be verified that the estimate for the intercept, 51.7957, is the average of the four block effects in Table 2.3.

From these examples, it can be seen that the orthogonal blocking of standard response surface designs is possible only for specific block sizes and numbers of center points. In practice, these parameters are mostly dictated by the experimental situation and orthogonally blocking the response surface design is in many cases impossible. Because the block sizes usually cannot be chosen at liberty, we assume that they are fixed.

In Section 1.5.1, we also pointed out that 2^m factorial designs can be blocked orthogonally at the expense of information on higher order interactions. As will be explained in Chapter 9, the levels of the higher order interactions are then used to assign the experimental runs to the blocks. This way of blocking is applicable when the block size is homogeneous and equal to a power of two.

2.3.6   Optimality of orthogonally blocked experiments

Assume that a design $\mathbf{X}$ or $\tilde{\mathbf{X}}$ is given and that the n rows have to be assigned to b blocks of size $k_1, k_2, \ldots, k_b$. If we arrange the n observations so that condition (2.39) is satisfied, the design is orthogonally blocked and the variance-covariance matrix of $\hat{\beta}$ is given by (2.42). If we arrange the n observations so that condition (2.39) is not satisfied, the design is not orthogonally blocked and we have that

$$\tilde{\mathbf{X}}_i'\mathbf{1}_{k_i} = \frac{k_i}{n}\tilde{\mathbf{X}}'\mathbf{1}_n + \boldsymbol{\delta}_i, \qquad i = 1, 2, \ldots, b, \tag{2.43}$$

where at least one $\boldsymbol{\delta}_i \neq \mathbf{0}_{p-1}$. Note that $\sum_{i=1}^{b} \boldsymbol{\delta}_i = \mathbf{0}_{p-1}$. As a matter of fact, summing (2.39) over all blocks yields

$$\sum_{i=1}^{b} \tilde{\mathbf{X}}_i'\mathbf{1}_{k_i} = \sum_{i=1}^{b} \frac{k_i}{n}\tilde{\mathbf{X}}'\mathbf{1}_n + \sum_{i=1}^{b} \boldsymbol{\delta}_i, \tag{2.44}$$

or

$$\tilde{\mathbf{X}}'\mathbf{1}_n = \frac{1}{n}\tilde{\mathbf{X}}'\mathbf{1}_n \sum_{i=1}^{b} k_i + \sum_{i=1}^{b} \boldsymbol{\delta}_i,$$
$$= \tilde{\mathbf{X}}'\mathbf{1}_n + \sum_{i=1}^{b} \boldsymbol{\delta}_i, \tag{2.45}$$

so that

$$\sum_{i=1}^{b} \boldsymbol{\delta}_i = \tilde{\mathbf{X}}'\mathbf{1}_n - \tilde{\mathbf{X}}'\mathbf{1}_n = \mathbf{0}_{p-1}. \tag{2.46}$$

The variance-covariance matrix of $\hat{\beta}$ can then be written as

$$\mathrm{var}(\hat{\beta}_{\mathrm{n.orth.}}) = \sigma_\varepsilon^2 \{\tilde{\mathbf{X}}'\tilde{\mathbf{X}} - \sum_{i=1}^{b} \frac{1}{k_i}\left(\frac{k_i}{n}\tilde{\mathbf{X}}'\mathbf{1}_n + \boldsymbol{\delta}_i\right)\left(\frac{k_i}{n}\tilde{\mathbf{X}}'\mathbf{1}_n + \boldsymbol{\delta}_i\right)'\}^{-1},$$

$$= \sigma_\varepsilon^2 \{\tilde{\mathbf{X}}'\tilde{\mathbf{X}} - \sum_{i=1}^{b} \frac{1}{k_i}\left(\frac{k_i}{n}\tilde{\mathbf{X}}'\mathbf{1}_n\right)\left(\frac{k_i}{n}\tilde{\mathbf{X}}'\mathbf{1}_n\right)' - \sum_{i=1}^{b} \frac{k_i}{nk_i}(\tilde{\mathbf{X}}'\mathbf{1}_n)\boldsymbol{\delta}_i'$$

$$- \sum_{i=1}^{b} \frac{k_i}{nk_i}\boldsymbol{\delta}_i(\tilde{\mathbf{X}}'\mathbf{1}_n)' - \sum_{i=1}^{b} \frac{1}{k_i}\boldsymbol{\delta}_i\boldsymbol{\delta}_i'\}^{-1},$$

$$= \sigma_\varepsilon^2 \{\tilde{\mathbf{X}}'\tilde{\mathbf{X}} - n^{-1}(\tilde{\mathbf{X}}'\mathbf{1}_n)(\tilde{\mathbf{X}}'\mathbf{1}_n)' - \sum_{i=1}^{b} \frac{1}{k_i}\boldsymbol{\delta}_i\boldsymbol{\delta}_i'\}^{-1},$$

$$= \sigma_\varepsilon^2 \{\tilde{\mathbf{X}}'\tilde{\mathbf{X}} - n^{-1}(\tilde{\mathbf{X}}'\mathbf{1}_n)(\tilde{\mathbf{X}}'\mathbf{1}_n)' - \boldsymbol{\Delta}'\boldsymbol{\Delta}\}^{-1}, \tag{2.47}$$

where $\boldsymbol{\Delta} = [\, \boldsymbol{\delta}_1/\sqrt{k_1} \quad \boldsymbol{\delta}_2/\sqrt{k_2} \quad \ldots \quad \boldsymbol{\delta}_b/\sqrt{k_b}]'$. The information matrix on the unknown β in the case of an orthogonal and a non-orthogonal design is given by the inverse of $\text{var}(\hat{\boldsymbol{\beta}}_{\text{orth.}})$ and $\text{var}(\hat{\boldsymbol{\beta}}_{\text{n.orth.}})$ respectively:

$$\mathbf{M}_{\text{orth.}} = \sigma_\varepsilon^{-2}\{\tilde{\mathbf{X}}'\tilde{\mathbf{X}} - n^{-1}(\tilde{\mathbf{X}}'\mathbf{1}_n)(\tilde{\mathbf{X}}'\mathbf{1}_n)'\}, \tag{2.48}$$

and

$$\mathbf{M}_{\text{n.orth.}} = \sigma_\varepsilon^{-2}\{\tilde{\mathbf{X}}'\tilde{\mathbf{X}} - n^{-1}(\tilde{\mathbf{X}}'\mathbf{1}_n)(\tilde{\mathbf{X}}'\mathbf{1}_n)' - \boldsymbol{\Delta}'\boldsymbol{\Delta}\}. \tag{2.49}$$

Both matrices are symmetric and positive definite. Now, the difference

$$\mathbf{M}_{\text{orth.}} - \mathbf{M}_{\text{n.orth.}} = \sigma_\varepsilon^{-2}\boldsymbol{\Delta}'\boldsymbol{\Delta} \tag{2.50}$$

is nonnegative definite. As a result, for any given $\mathbf{X}$, orthogonally blocked designs will be better for the estimation of the model parameters apart from the intercept and the block effects, that is $\tilde{\boldsymbol{\beta}}$, than designs that are not blocked orthogonally. As was already indicated in Section 1.4.2, this result is valid for all generalized optimality criteria, for example for the $\mathcal{D}$-optimality criterion. The $\mathcal{D}$-optimality of orthogonal blocking can also be shown using Corollary 18.1.8 of Harville (1997), which states that, for any square symmetric positive definite matrix $\mathbf{A}$ and for any symmetric matrix $\mathbf{C}$, $\mathbf{C} \neq \mathbf{A}$, so that $\mathbf{C} - \mathbf{A}$ is nonnegative definite,

$$|\mathbf{C}| > |\mathbf{A}|.$$

As a result,

$$|\mathbf{M}_{\text{orth.}}| > |\mathbf{M}_{\text{n.orth.}}|. \tag{2.51}$$

We have shown that orthogonal blocking is optimal when the interest is in estimating $\tilde{\boldsymbol{\beta}}$, that is when the interest is only in estimating the factor effects and not the intercept and the block effects. Nevertheless, orthogonal blocking also turns out to be a $\mathcal{D}$-optimal strategy when the interest is in estimating $\boldsymbol{\theta} = [\ \tilde{\boldsymbol{\beta}}'\ \ \boldsymbol{\tau}'\ ]'$. This is because maximizing the determinant of (2.31) is equivalent to maximizing the determinant of (2.37). In order to see this, apply Theorem 13.3.8 of Harville (1997) to find that

$$\begin{vmatrix} \tilde{\mathbf{X}}'\tilde{\mathbf{X}} & \tilde{\mathbf{X}}'\mathbf{Z} \\ \mathbf{Z}'\tilde{\mathbf{X}} & \mathbf{Z}'\mathbf{Z} \end{vmatrix} = |\mathbf{Z}'\mathbf{Z}|\,|\tilde{\mathbf{X}}'\tilde{\mathbf{X}} - \tilde{\mathbf{X}}'\mathbf{Z}(\mathbf{Z}'\mathbf{Z})^{-1}\mathbf{Z}'\tilde{\mathbf{X}}|,$$

$$= \prod_{i=1}^{b} k_i \left| \tilde{\mathbf{X}}'\tilde{\mathbf{X}} - \sum_{i=1}^{b} \frac{1}{k_i}(\tilde{\mathbf{X}}_i'\mathbf{1}_{k_i})(\tilde{\mathbf{X}}_i'\mathbf{1}_{k_i})' \right|.$$

Now, $\prod_{i=1}^{b} k_i$ is constant for all possible designs, so that the determinant at the left hand side of the equation is maximal when that at the right hand side is maximal and vice versa. In other words, the $\mathcal{D}_s$- or $\mathcal{D}_{\tilde{\beta}}$-optimal design for the estimation of $\tilde{\boldsymbol{\beta}}$ is equivalent to the $\mathcal{D}$-optimal design for estimating both $\tilde{\boldsymbol{\beta}}$ and $\boldsymbol{\tau}$.

2.3.7 Constructing D-optimal blocked experiments

Orthogonally blocked designs

It is clear that orthogonal blocking of an experiment is the best option for a given $\mathbf{X}$ in terms of $\mathcal{D}$-optimality. The variance-covariance matrix of the least squares estimator as well as the information matrix on the unknown $\tilde{\beta}$ are identical to those of a model without block effects. If we let $\mathbf{X}$ be a $\mathcal{D}$-optimal design for the model without block effects, that is model (1.3), then assigning the corresponding design points to the blocks so that the resulting experiment is orthogonally blocked yields a $\mathcal{D}$-optimal block design.

General case

Unfortunately, the number of blocks and the block sizes often make it impossible to block a given design orthogonally. Algorithms to construct $\mathcal{D}$-optimal blocking designs for a response surface model with fixed block effects have been proposed by Atkinson and Donev (1989) and by Cook and Nachtsheim (1989). Both algorithms will be described in Chapter 4.

2.3.8 Product designs

Kurotschka (1981) shows that $\mathcal{D}$- and $\mathcal{A}$-optimal continuous designs for the fixed effects model, that is the linear model (2.11) with distributional assumption (2.13), consist of replications of the optimal design for the uncorrelated model. In other words, the blocks of the optimal experiment are identical. Therefore, their observations can be denoted by a single measure ν. Since the experiment consists of b blocks with weight $1/b$, the design can be represented by

$$\xi = \kappa \times \nu,$$

where κ is the measure of which all elements are equal to $1/b$. Since this expression shows that the optimal design derived by Kurotschka (1981) can be written as a product of two measures, these designs are called product designs.

These optimal continuous designs can be used for the construction of optimal discrete designs when the weights of the measure ν, say w_i, are rational numbers and when the block sizes k_i are a multiple of the smallest common denominator of the weights w_i. As an illustration, consider an experiment with two blocks of six observations in order to estimate a quadratic model in one explanatory variable. Thus, $b = 2$ and $k_1 = k_2 = 6$. The $\mathcal{D}$-optimal continuous design for estimating a second order polynomial has equal weight $w_i = 1/3$ on three design points: -1, 0 and 1. The block sizes are obviously multiples of 3, which is a multiple of the smallest common denominator of the w_i. As a result, the $\mathcal{D}$-optimal design for quadratic regression with two blocks of size six a has two identical blocks with two observations in the points -1, 0 and 1.

The designs described here are orthogonal designs because the average row of $\mathbf{X}_i$ is the same for all blocks. Moreover, the optimal design points are given by the optimal design for a model without block effects. Therefore, the product designs can be seen as a special case of the $\mathcal{D}$-optimal orthogonally blocked designs described in the previous section.

2.3.9   The efficiency of blocking

Blocked experiments are used because they allow a more efficient estimation of the factor effects. This is because part of the total variance in the data can be attributed to the blocks, so that the residual variance is smaller. Let us denote by σ_r^2 the residual variance of a completely randomized experiment and by σ_b^2 the residual variance of a blocked experiment. Suppose that the orthogonally blocked central composite design of Table 2.1 is used to estimate a full second order model in three variables:

$$y = \beta_0 + \sum_{i=1}^{3} \beta_i x_i + \sum_{i=1}^{3}\sum_{j=i}^{3} \beta_{ij} x_i x_j.$$

The variance-covariance matrix of the parameter estimate $\hat{\tilde{\beta}}$ is then given by

$$\sigma_b^2 \begin{bmatrix} 0.075\ \mathbf{I}_3 & \mathbf{0}'_{3\times3} & \mathbf{0}_3 & \mathbf{0}_3 & \mathbf{0}_3 \\ \mathbf{0}'_{3\times3} & 0.125\ \mathbf{I}_3 & \mathbf{0}_3 & \mathbf{0}_3 & \mathbf{0}_3 \\ \mathbf{0}'_3 & \mathbf{0}'_3 & 0.076 & 0.005 & 0.005 \\ \mathbf{0}'_3 & \mathbf{0}'_3 & 0.005 & 0.076 & 0.005 \\ \mathbf{0}'_3 & \mathbf{0}'_3 & 0.005 & 0.005 & 0.076 \end{bmatrix}. \qquad (2.52)$$

The information matrix on the unknown $\tilde{\beta}$ is the inverse of this variance-covariance matrix and it can be verified that its determinant amounts to $2.8368\ \text{E9}/\sigma_b^{18}$. Let us denote this value by $\mathcal{D}_b$. When the design points in Table 2.1 are used in a completely randomized instead of an orthogonally blocked experiment, the variance-covariance matrix of the parameter estimate $\hat{\tilde{\beta}}$ is equal to

$$\sigma_r^2 \begin{bmatrix} 0.075\ \mathbf{I}_3 & \mathbf{0}'_{3\times3} & \mathbf{0}_3 & \mathbf{0}_3 & \mathbf{0}_3 \\ \mathbf{0}'_{3\times3} & 0.125\ \mathbf{I}_3 & \mathbf{0}_3 & \mathbf{0}_3 & \mathbf{0}_3 \\ \mathbf{0}'_3 & \mathbf{0}'_3 & 0.076 & 0.005 & 0.005 \\ \mathbf{0}'_3 & \mathbf{0}'_3 & 0.005 & 0.076 & 0.005 \\ \mathbf{0}'_3 & \mathbf{0}'_3 & 0.005 & 0.005 & 0.076 \end{bmatrix}. \qquad (2.53)$$

The information matrix on the unknown $\tilde{\beta}$ is the inverse of this variance-covariance matrix and it can be verified that its determinant amounts to

2.8368 E9$/\sigma_r^{18}$. Let us denote this value by $\mathcal{D}_r$. The relative $\mathcal{D}$-efficiency of both design options is calculated as

$$\left(\frac{\mathcal{D}_b}{\mathcal{D}_r}\right)^{1/9} = \left(\frac{2.8368\ \mathrm{E9}/\sigma_b^{18}}{2.8368\ \mathrm{E9}/\sigma_r^{18}}\right)^{1/9} = \frac{\sigma_r^2}{\sigma_b^2}. \qquad (2.54)$$

Since σ_b^2 is always smaller than σ_r^2, the relative $\mathcal{D}$-efficiency of blocking is always greater than one. The relative $\mathcal{D}$-efficiency derived here is identical to the definition of the relative efficiency of randomized complete block designs to completely randomized designs given in Neter, Kutner, Nachtsheim and Wasserman (1996). A randomized complete block design is a balanced block design (see Section 1.6.2) where each treatment appears exactly once in each block. Since all blocks are identical in a randomized complete block design, it is an orthogonally blocked design.

The relative $\mathcal{A}$-efficiency of the orthogonally blocked central composite design and its completely randomized counterpart is obtained by comparing the traces of the matrices (2.52) and (2.53). As for the relative $\mathcal{D}$-efficiency, we obtain

$$\frac{0.82716\ \sigma_r^2}{0.82716\ \sigma_b^2} = \frac{\sigma_r^2}{\sigma_b^2}.$$

Forsaking the principle of orthogonal blocking leads to a loss in efficiency. In order to see this, suppose that the first two blocks of Table 2.1 are merged. The variance-covariance matrix of the parameter estimate $\hat{\boldsymbol{\beta}}$ is then given by

$$\sigma_b^2 \begin{bmatrix} 0.075\ \mathbf{I}_3 & \mathbf{0}'_{3\times 3} & \mathbf{0}_3 & \mathbf{0}_3 & \mathbf{0}_3 \\ \mathbf{0}'_{3\times 3} & 0.125\ \mathbf{I}_3 & \mathbf{0}_3 & \mathbf{0}_3 & \mathbf{0}_3 \\ \mathbf{0}'_3 & \mathbf{0}'_3 & 0.079 & 0.008 & 0.008 \\ \mathbf{0}'_3 & \mathbf{0}'_3 & 0.008 & 0.079 & 0.008 \\ \mathbf{0}'_3 & \mathbf{0}'_3 & 0.008 & 0.008 & 0.079 \end{bmatrix}. \qquad (2.55)$$

The corresponding $\mathcal{D}$- and $\mathcal{A}$-criterion values of this design option amount to 2.5750 E9$/\sigma_b^{18}$ and $0.83596\sigma_b^2$. As a result the relative $\mathcal{D}$-efficiency of blocking is then

$$\left(\frac{\mathcal{D}_b}{\mathcal{D}_r}\right)^{1/9} = \left(\frac{2.5750\ \mathrm{E9}/\sigma_b^{18}}{2.8368\ \mathrm{E9}/\sigma_r^{18}}\right)^{1/9} = 0.9893\frac{\sigma_r^2}{\sigma_b^2} < \frac{\sigma_r^2}{\sigma_b^2},$$

while the relative $\mathcal{A}$-efficiency of blocking is

$$\frac{0.82716\ \sigma_r^2}{0.83596\ \sigma_b^2} = 0.9895\frac{\sigma_r^2}{\sigma_b^2}.$$

As a result, the relative efficiency of blocking depends on the design criterion when the design is not blocked orthogonally.

3
Compound Symmetric Error Structure

The experiments considered in the sequel of this book possess a compound symmetric error structure. This is due to the fact that the experiments, which fall into the category of bi-randomization designs or two-stratum designs, suffer from a restricted randomization. The category of bi-randomization designs or two-stratum designs is a subset of the so-called multi-stratum designs. In this chapter, we provide the reader with a number of examples and give a general description of this type of design. Next, we show how to write the information matrix of a bi-randomization experiment as a sum of outer products without using Cholesky decomposition. This is important for the design construction algorithms in the next chapters. Finally, we will demonstrate that the asymptotic information matrix of a bi-randomization experiment is a reliable approximation to the finite sample information matrix.

3.1 Restricted randomization

One of the basic principles in experimental design is randomization. In order to carry out an experiment, the factor level combinations of the experimental runs are usually randomized. The main reason for this randomization is to make sure that systematic effects do not influence the results of the experiment. Randomization thus ensures that only the pure effects of the experimental variables are measured and that they are not confounded with extraneous factors which are not under the control of the experimenters.

When a complete randomization is used, it is implicitly assumed that the n experimental observations can be made in homogeneous circumstances.

However, the randomization is often restricted because some of the experimental factors are hard to change or hard to control. Another reason for a restricted randomization is that not all observations can be made in homogeneous circumstances. The following examples will illustrate the necessity of restricted randomization in some experimental situations.

Example 1. Letsinger, Myers and Lentner (1996) describe an experiment from the chemical industry in which the effect of five process variables, called temperature 1, temperature 2, humidity 1, humidity 2 and pressure, on a certain quality characteristic was investigated. The response surface design used was a modified central composite design. However, the different factor level combinations of the design were not carried out in a completely random order because the levels of the factor temperature 1 and pressure were hard to change. Instead, all the runs with the same level for these two factors were grouped and all runs within one group were carried out immediately after each other. In doing so, it was much easier to conduct the experiment because the levels of the hard-to-change factors were changed as little as possible.

It is clear that this experiment suffers from a restricted randomization since the order of the experimental runs is to some extent dictated by the levels of the experimental variables. In order to avoid that the experimental results are influenced by systematic effects, the different groups of observations are executed in a random order. In addition, the observations within one group need to be randomized as well. As a result, two different randomization procedures are performed: one randomization at the group level and one randomization at the level of the individual observations. Therefore, this type of experiment is referred to as a bi-randomization experiment. In part of the design literature, the two levels of randomization, the group level and the level of the individual observations, are called strata. Therefore, bi-randomization designs are also referred to as two-stratum designs. The group level is the higher level and it is called the first stratum U_1. The level of the individual observations is the lower level and is called the second stratum U_2. The two-stratum experiment is then denoted by $U_1|U_2$. The first stratum U_1 consists of so-called larger experimental units, which are divided in smaller experimental units in the second stratum U_2. In the first stratum, the groups of observations are randomly assigned to the larger experimental units. In the second stratum, the individual observations are assigned to the smaller experimental units.

Example 2. Trinca and Gilmour (2001) describe an experiment to investigate the effect of five factors on protein extraction. The factors were

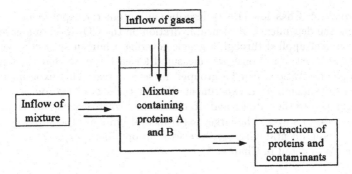

Figure 3.1: Protein extraction process.

the feed position for the inflow of the mixture, the feed flow rate, the gas flow rate, the concentration of protein A and the concentration of protein B. A schematic representation of the extraction process is given in Figure 3.1. Three levels were used for each factor. Since setting the feed position involved taking apart and reassembling the equipment and was time-consuming, it was decided that the feed position should only be changed after one day of experimentation. Thanks to this policy, two experimental runs instead of one could be performed on one single day. Therefore, 42 experimental runs could be carried out in 21 days. It is clear that this experiment strongly resembles that of Example 1. Again, there are two strata, the days and the individual runs, and two randomization procedures. In Chapter 6, this example will be revisited.

Example 3. Often, the material used in experiments is taken from batches and, in many cases, the quality of the material varies considerably from one batch to another. Observations originating from the same batch then form a group of possibly correlated observations. In order to perform the experiment, a number of batches is randomly selected from a population of batches. Next, each batch is divided in smaller units of experimental material. As a result, the experiment also has two different levels. At the higher level, the batches correspond to a group of observations. At the lower level, each unit of material corresponds to a single experimental run. Again, there are two randomization procedures involved. Firstly, the batches are assigned to the groups of observations. Secondly, within one group, the smaller units of experimental material are assigned to the individual runs. Khuri (1992) gives an example from the steel industry where 12 batches of steel were randomly selected from the warehouse. Per batch, 10, 12 or 13 observations were carried out. The purpose of the experiment was to study the effects of curing time and temperature on the shear strength of a steel bonding.

Example 4. Chasalow (1992) describes an optometry experiment for exploring the dependence of corneal hydration on the CO_2 level in a gaseous environment applied through a goggle covering a human subject's eye. In the experiment, a response was measured for each eye, so that the experimental observations can be grouped in sets of two. This experiment is also a bi-randomization experiment. Firstly, the sets of two observations are assigned to the subject and, secondly, the two treatments are assigned to the subject's eyes. The larger experimental units are the subjects. The smaller experimental units are the eyes. This optometry experiment will be examined in detail in Chapter 5.

Example 5. Bisgaard and Steinberg (1997) use an example from Taguchi (1989) to illustrate how prototype experiments are designed. The purpose of the experiment was to reduce the CO content of exhaust gas. Seven hard-to-change factors, A, B, C, D, E, F and G, each possessing two levels, were studied, along with three driving modes R_1, R_2 and R_3. Due to cost considerations, only 8 of the 2^7 combinations of the hard-to-change factor level combinations were used in the experiment. The three driving modes correspond to increasing numbers of rotations per minute. Completely randomizing the entire experiment was impossible because this would imply that 8×3 prototype engines would have to be built, that is one for each experimental run. However, in order to save costs, only eight prototype engines were developed and each prototype was used under the three driving modes. It is clear that the prototype experiment is also a two-stratum or a bi-randomization experiment. The first stratum is the prototype level, whereas the second stratum is the individual run.

Example 6. Cornell and Gorman (1984) describe an experiment to investigate the texture of sandwich patties as a function of six experimental factors, three of which are mixture variables and three of which are process variables. The patties were a mixture of three fish species: mullet, sheepshead and croaker. Seven different blends of the three species were examined. The process variables suspected to influence the texture were cooking temperature, cooking time and deep fat frying time. Each process variable had two levels, yielding eight different processing conditions. An easy way to perform the experiment is to select one of the seven blends at random and to prepare eight patties using this blend. The eight processing conditions are then performed as a group, one combination for each patty. When these runs are completed, a second blend is chosen, eight patties are made up, and the eight processing conditions are carried out. This procedure is repeated until the seven blends have been used. The fish patty experiment is an example of a mixture experiment with process variables. As pointed out in Cornell (1988), this type of experiment is often conducted in a bi-randomization format.

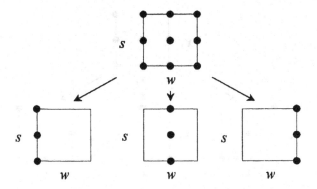

Figure 3.2: The structure of a split-plot experiment.

Within the class of bi-randomization or two-stratum designs, at least two subclasses can be identified: the split-plot designs and the block designs. In a split-plot design, some of the factor levels are connected to the groups of experimental runs. Consider, for instance, the chemical experiment from Example 1. In this experiment, one combination of the temperature 1 and the pressure level corresponds to each group of observations. It is therefore said that the factors temperature 1 and pressure are applied to the first stratum U_1. The remaining factors, temperature 2, humidity 1 and 2, are applied to the second stratum U_2. The experiment from Example 2 is also a split-plot experiment. As a matter of fact, the factor feed position is applied to the first stratum, the days, while the remaining four factors are applied to the second stratum. The prototype experiment is another illustration of a split-plot design because the eight combinations of the hard-to-change factors A to G are connected with the first stratum, the prototype. Finally, the last example is a split-plot experiment too because each of the seven mixtures corresponds to a group of eight observations. In the split-plot literature, the factors applied to the first stratum are referred to as whole plot factors. Typically, it is expensive and/or time-consuming to change the levels of these factors. Therefore, they are also referred to as hard-to-change factors. The factors applied to the second stratum are called sub-plot factors. Usually, these factors are easier to change. By analogy, the larger experimental units from the first stratum are called the whole plots, whereas the smaller experimental units from the second stratum are called the sub-plots. This terminology, originally used in agricultural experiments, is derived from the word plot, which signifies a piece of land. The structure of a simple split-plot experiment with nine observations, one whole plot variable w and one sub-plot variable s is visualized in Figure 3.2.

The experiments described in Examples 3 and 4 are block designs. In contrast with split-plot experiments, there is no connection between the groups

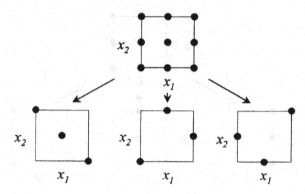

Figure 3.3: The structure of a blocked experiment.

of runs and the factor levels. In the optometry experiment of Example 4, the only experimental factor, the CO_2 level, is applied to the second stratum, the eyes. Hence, no factor is applied to the first stratum. In Example 3, no factor levels are associated with the different batches of steel and all factors are applied to the second stratum. In a block design, the larger experimental units are referred to as blocks instead of whole plots. The problem of designing a blocked experiment for a particular situation is totally different from designing a split-plot experiment because no restriction is imposed on the factor levels of the experimental runs belonging to the same block. This is illustrated in Figure 3.3, where a simple blocked experiment in two experimental factors x_1 and x_2 is displayed.

It is clear that the terms whole plot and block are synonyms since they refer to a group of experimental observations. However, the term whole plot is only used in the split-plot literature, that is when some factors are applied to the larger experimental units. On the contrary, the term block is only used when no factors are applied to the larger experimental units. In the sequel of this text, we will also use the terms whole plot and block in this fashion. We will use the term group for statements that are valid for both split-plot experiments and blocked experiments.

It should also be pointed out that, in some experimental situations, the number of groups and the group size is fixed whereas, in other situations, the number of groups and the group size can be chosen freely. In the optometry experiment, for example, the group size is two, simply because each human has two eyes. In the protein experiment of Example 2, the group size is two because this is the maximum number of observations that can be made on a single day. In the chemical experiment of Example 1, the group size is obviously not dictated by the experimental situation: some of

Table 3.1: Design problems considered in this book.

	Block Designs	Split-Plot Designs
Number of groups and group sizes fixed in advance	Optimal Designs for the Random Block Effects Model (Chapter 4) Optimal Designs for Quadratic Regression on one Variable and Blocks of Size Two (Chapter 5) 2^m and 2^{m-f} Blocked Designs (Section 9.2)	Constrained Split-Plot Designs (Chapter 6) 2^{m-f} Split-Plot Designs (Section 9.3)
Number of groups and group sizes optimally chosen	Optimal Number of Blocks and Block Sizes (Section 4.10)	Optimal Split-Plot Designs in the Presence of Hard-to-Change Factors (Chapter 7) Optimal Split-Plot Designs (Chapter 8)

the groups contain only one run, whereas other groups contain four runs and one group even has size seven.

An overview of the design problems tackled in this book is given in Table 3.1. In the next chapter, we will concentrate on the design of experiments for estimating the random block effects model. This model was introduced in Section 2.3.1. In most practical instances, the number of groups (blocks) is determined by a sort of budget or time constraint. Moreover, the block size is dictated by the experimental situation. This is also the case in the optometry experiment introduced in Example 4. This experiment is examined in detail in Chapter 5. In Section 9.2, we show how 2^m and 2^{m-f} designs can be arranged in blocks of size 2^q. The experimental situation in which the number of blocks and the block size can be chosen at liberty only receives attention in Section 4.10. The design of split-plot experiments is treated in Chapters 6, 7 and 8, and in Section 9.3.

In Chapter 6 and Section 9.3, we will consider the design of split-plot experiments where the number of groups (whole plots) and the group sizes (whole plot sizes) are fixed in advance. An illustration of this design problem is the protein experiment described in Example 2: 21 days were available and two observations could be made per day. In Chapter 7, another kind of restriction is imposed on the design: only one group of observations is connected to a given combination of the hard-to-change or whole plot factor levels. This design problem is met in the prototype experiment of Example

5: for each combination of the factors A to G, only one prototype is built. Of course, this restriction is inspired by cost considerations. Another experiment subject to this restriction is the experiment of Example 1: all observations possessing the same levels for the factors temperature 1 and pressure are carried out successively and therefore form a group of possibly correlated observations. Finally, in Chapter 8 no restriction whatsoever is imposed on the design. The split-plot designs derived in that chapter are especially useful when none of the experimental factors are hard to change, that is when the completely randomized experiment is an option. We will show that a completely randomized experiment is often outperformed by a split-plot experiment.

Despite the fact that they frequently occur in industrial experimentation, the design problems considered in this book have received relatively little attention and no algorithms exist to design optimal tailor-made bi-randomization designs. Therefore, researchers often fall back on standard response surface designs, even though these designs are not very flexible and usually do not match the experimental situation. Rather than developing a designed experiment for a particular problem, the situation is adapted to the experimental design available. It is the purpose of this book to develop design construction algorithms to help the experimenter in designing an optimal bi-randomization experiment, taking into account the constraints dictated by the experimental situation.

In each of the bi-randomization or two-stratum experiments, two randomization procedures are used. Therefore, two error terms will occur in the statistical model corresponding to the experiment and the error structure of the experiments is compound symmetric. This is shown in the next section. In Section 3.3, we show how to analyze a bi-randomization experiment, and in Section 3.4, the compound symmetric error structure is exploited to obtain a computationally attractive expression for the information matrix of a bi-randomization design.

3.2   Model

The statistical model corresponding to a bi-randomization or a two-stratum experiment with n observations arranged in b groups of size k_i $(i = 1, 2, \ldots, b)$ is a linear mixed model of the form

$$y = X\beta + Z\gamma + \varepsilon, \tag{3.1}$$

where y is the vector of observed responses, X is the (extended) design matrix containing the settings of the experimental variables in the n experimental runs, β is the p-dimensional vector of factor effects, Z is a matrix of zeroes and ones assigning the observations to the groups, γ is the b-dimensional vector of group effects and $\boldsymbol{\varepsilon}$ is the vector containing the random errors. It is assumed that there is no interaction between $\boldsymbol{\beta}$ and $\boldsymbol{\gamma}$.

The model contains two random components, that is one for each randomization. The first random component is $\boldsymbol{\gamma}$ and the second one is $\boldsymbol{\varepsilon}$. It is assumed that

$$\mathrm{E}(\boldsymbol{\varepsilon}) = \mathbf{0}_n \text{ and } \mathrm{Cov}(\boldsymbol{\varepsilon}) = \sigma_\varepsilon^2 \mathbf{I}_n, \tag{3.2}$$

$$\mathrm{E}(\boldsymbol{\gamma}) = \mathbf{0}_b \text{ and } \mathrm{Cov}(\boldsymbol{\gamma}) = \sigma_\gamma^2 \mathbf{I}_b, \tag{3.3}$$

$$\text{and } \mathrm{Cov}(\boldsymbol{\gamma}, \boldsymbol{\varepsilon}) = \mathbf{0}_{b \times n}. \tag{3.4}$$

Under these assumptions, the variance-covariance matrix of the observations $\mathrm{Cov}(\mathbf{y})$ can be written as

$$\mathbf{V} = \sigma_\varepsilon^2 \mathbf{I}_n + \sigma_\gamma^2 \mathbf{Z}\mathbf{Z}'. \tag{3.5}$$

Suppose the entries of $\mathbf{y}$ are arranged per group, then

$$\mathbf{V} = \begin{bmatrix} \mathbf{V}_1 & \mathbf{0} & \cdots & \mathbf{0} \\ \mathbf{0} & \mathbf{V}_2 & \cdots & \mathbf{0} \\ \vdots & & \ddots & \vdots \\ \mathbf{0} & \cdots & \mathbf{0} & \mathbf{V}_w \end{bmatrix}, \tag{3.6}$$

where

$$\begin{aligned} \mathbf{V}_i &= \sigma_\varepsilon^2 \mathbf{I}_{k_i} + \sigma_\gamma^2 \mathbf{1}_{k_i} \mathbf{1}'_{k_i}, \\ &= \begin{bmatrix} \sigma_\varepsilon^2 + \sigma_\gamma^2 & \sigma_\gamma^2 & \cdots & \sigma_\gamma^2 \\ \sigma_\gamma^2 & \sigma_\varepsilon^2 + \sigma_\gamma^2 & \cdots & \sigma_\gamma^2 \\ \vdots & & \ddots & \vdots \\ \sigma_\gamma^2 & \sigma_\gamma^2 & \cdots & \sigma_\varepsilon^2 + \sigma_\gamma^2 \end{bmatrix}. \end{aligned} \tag{3.7}$$

As a result, the variance-covariance matrix $\mathbf{V}_i$ of all observations within one group is compound symmetric: the main diagonal of the matrix contains the constant variances of the observations, while the off-diagonal elements are constant covariances. However, we prefer rewriting this expression as

$$\begin{aligned} \mathbf{V}_i &= \sigma_\varepsilon^2 (\mathbf{I}_{k_i \times k_i} + \eta \mathbf{1}_{k_i} \mathbf{1}'_{k_i}), \\ &= \sigma_\varepsilon^2 \begin{bmatrix} 1+\eta & \eta & \cdots & \eta \\ \eta & 1+\eta & \cdots & \eta \\ \vdots & & \ddots & \vdots \\ \eta & \eta & \cdots & 1+\eta \end{bmatrix}, \end{aligned} \tag{3.8}$$

where $\eta = \sigma_\gamma^2 / \sigma_\varepsilon^2$ is a measure for the extent to which observations within the same group are correlated. This is because, for the purpose of design construction, the ratio of both variance components matters, but not their absolute magnitude. In this book, we will refer to η as the degree of correlation. Since both σ_ε^2 and σ_γ^2 are positive numbers, η is also positive. The larger η, the more the observations within one group are correlated. The smaller η, the less they are correlated.

While any two observations in a given group are correlated in advance of the experimental runs, once a block has been selected, the additive model (3.1) assumes that the observations are independent. As a matter of fact, the only remaining variation in an observation then is the random error ε and it is assumed that the random errors are independent. In the next sections, we examine the consequences of this particular variance-covariance matrix.

3.3   Analysis

The statistical data analysis of bi-randomization experiments differs from that of completely randomized experiments. When the random error terms as well as the group effects are normally distributed, the maximum likelihood estimate of the unknown model parameter β in (3.1) is the generalized least squares (GLS) estimate instead of the ordinary least squares (OLS) estimate. As a result, the unknown model parameters β are estimated by

$$\hat{\beta} = (\mathbf{X}'\mathbf{V}^{-1}\mathbf{X})^{-1}\mathbf{X}'\mathbf{V}^{-1}\mathbf{y}, \tag{3.9}$$

and the variance-covariance matrix of the estimators is given by

$$\mathrm{var}(\hat{\beta}) = (\mathbf{X}'\mathbf{V}^{-1}\mathbf{X})^{-1}. \tag{3.10}$$

Usually, however, the variances σ_γ^2 and σ_ε^2 are not known and therefore, (3.9) and (3.10) cannot be used directly. Instead the variance components σ_γ^2 and σ_ε^2 are estimated, and the estimates $\hat{\sigma}_\gamma^2$ and $\hat{\sigma}_\varepsilon^2$ are substituted in the GLS estimator (3.9), yielding the so-called feasible GLS estimator

$$\hat{\beta} = (\mathbf{X}'\hat{\mathbf{V}}^{-1}\mathbf{X})^{-1}\mathbf{X}'\hat{\mathbf{V}}^{-1}\mathbf{y}, \tag{3.11}$$

where

$$\hat{\mathbf{V}} = \hat{\sigma}_\varepsilon^2 \mathbf{I}_n + \hat{\sigma}_\gamma^2 \mathbf{Z}\mathbf{Z}'.$$

In that case, the variance-covariance matrix (3.10) can be approximated by

$$\mathrm{var}(\hat{\beta}) = (\mathbf{X}'\hat{\mathbf{V}}^{-1}\mathbf{X})^{-1}. \tag{3.12}$$

Estimates for the variance components σ_γ^2 and σ_ε^2 are thoroughly described by Letsinger et al. (1996). They recommend restricted maximum likelihood (REML) for error variance estimation because it performs well for various values of η and because it is also a good estimation option when smaller designs and near full second order models are used. REML is the default estimation method in the SAS procedure PROC MIXED for estimating linear mixed models. The diagonal elements of (3.12) are used as the estimated variances for the parameter estimates. This approach, however, leads to an underestimation of the variances because it ignores the stochastic nature of $\hat{\sigma}_\gamma^2$ and $\hat{\sigma}_\varepsilon^2$. This was shown by Kackar and Harville (1984). Gilmour, Thompson and Cullis (1995) and Kenward and Roger (1997) present methods to tackle this problem.

The information matrix on the unknown fixed model parameters β is given by the inverse of the variance-covariance matrix and is denoted by

$$\mathbf{M} = \mathbf{X}'\mathbf{V}^{-1}\mathbf{X}. \tag{3.13}$$

Improper analysis of a block design as if it was a completely randomized experiment leads to a less powerful analysis. The unexplained variance of the experiment is larger and therefore it will be more difficult to find significant effects. The risks of improper analysis of split-plot experiments receive detailed attention in Section 6.3.

3.3.1   The analysis of a blocked experiment

An example of a blocked experiment was carried out in the Department of Food Science and Technology at the University of Reading in 1995. The experiment was introduced in the literature by Trinca and Gilmour (2000) and Gilmour and Trinca (2000). The factors investigated in the experiment were the feed flow rate (expressed in kg/h), the initial moisture content (in %) and the screw speed (in rotations per minute) of a mixing process for pastry dough. The purpose of the experimenter was to understand how the various properties of the dough depend on these three factors and to develop an overall control scheme based on the experimental results. In one day, it was possible to perform only four runs, and since an important day to day variation was expected, blocks of size four were needed to conduct the experiment. It was decided that seven days of experimentation were affordable, so that 28 runs could be performed. Nine response variables were measured, two of which will be studied here. Each response measured the light reflectance in a different band of the spectrum. As a result, the responses served as an indication of how the color of the dough is affected as the factor levels are changed. The data of the experiment are given in Table 3.2. The table contains both the uncoded and coded factor levels.

A full second order model in the three explanatory variables was used to explain the behavior of the two responses in Table 3.2. The jth observation within the ith block can then be written as

$$y_{ij} = \beta_0 + \sum_{i=1}^{3}\beta_i x_i + \sum_{i=1}^{3}\sum_{j=i+1}^{3}\beta_{ij} x_i x_j + \sum_{i=1}^{3}\beta_{ii} x_i^2 + \gamma_i + \varepsilon_{ij}.$$

Table 3.2: Data for the pastry dough mixing experiment.

Block	Flow rate	Moisture content	Screw speed	x_1	x_2	x_3	y_1	y_2
1	30.0	18	300	-1	-1	-1	12.92	77.00
	30.0	24	400	-1	1	1	13.91	76.73
	45.0	18	400	1	-1	1	11.66	78.38
	45.0	24	300	1	1	-1	14.48	77.19
2	30.0	18	400	-1	-1	1	10.76	78.68
	30.0	24	300	-1	1	-1	14.41	77.74
	45.0	18	300	1	-1	-1	12.27	76.9
	45.0	24	400	1	1	1	12.13	77.24
3	30.0	24	300	-1	1	-1	14.22	76.79
	37.5	18	350	0	-1	0	12.35	76.75
	45.0	21	350	1	0	0	13.50	76.70
	37.5	21	400	0	0	1	12.54	77.64
4	45.0	18	400	1	-1	1	10.55	78.07
	30.0	21	350	-1	0	0	13.33	76.99
	37.5	24	350	0	1	0	13.84	76.34
	37.5	21	300	0	0	-1	14.19	76.72
5	30.0	18	300	-1	-1	-1	11.46	77.89
	45.0	24	400	1	1	1	11.32	78.10
	37.5	21	350	0	0	0	11.93	77.31
	37.5	21	350	0	0	0	11.63	77.91
6	30.0	18	400	-1	-1	1	12.20	77.74
	45.0	24	300	1	1	-1	14.78	77.19
	37.5	21	350	0	0	0	14.94	76.97
	37.5	21	350	0	0	0	14.61	76.97
7	30.0	24	400	-1	1	1	12.17	76.93
	45.0	18	300	1	-1	-1	11.28	77.96
	37.5	21	350	0	0	0	11.85	77.51
	37.5	21	350	0	0	0	11.64	77.38

Estimates of all factor effects and the variance components σ_ε^2 and σ_γ^2 can be computed using PROC MIXED in SAS. The following commands are needed:

```
data trgi;
infile 'pastry.dat';
input block x1 x2 x3 y1 y2;
x1x2=x1*x2;
x1x3=x1*x3;
x2x3=x2*x3;
x11=x1*x1;
x22=x2*x2;
```

Table 3.3: Coefficient estimates, standard errors, t-statistics and p-values for response y_1 in the pastry dough mixing experiment.

Term	Estimate	St. error	t-value	p-value
Intercept	13.1960	0.3910	33.75	$< .0001$
x_1	-0.1894	0.0734	-2.58	0.0240
x_2	0.8783	0.0734	11.97	$< .0001$
x_3	-0.7094	0.0734	-9.67	$< .0001$
$x_1 x_2$	-0.1844	0.0879	-2.10	0.0578
$x_1 x_3$	-0.0648	0.0879	-0.74	0.4750
$x_2 x_3$	0.1608	0.0879	1.83	0.0922
x_1^2	-0.1103	0.1877	-0.59	0.5676
x_2^2	-0.4303	0.1877	-2.29	0.0407
x_3^2	-0.1603	0.1877	-0.85	0.4097

```
x33=x3*x3;
proc mixed;
class block;
model y1 = x1 x2 x3 x1x2 x1x3 x2x3 x11 x22 x33 / solution;
random block;
proc mixed;
class block;
model y2 = x1 x2 x3 x1x2 x1x3 x2x3 x11 x22 x33 / solution;
random block;
run;
```

The estimated coefficients, as well as the estimated standard errors, the t-statistics and the p-values produced by SAS are displayed in Table 3.3 and Table 3.4. For the computations, the coded factor levels were used. Table 3.3 contains the results for the response y_1, while Table 3.4 contains the results for y_2. Twelve degrees of freedom are associated with every coefficient, except for the intercept with which six degrees of freedom are associated. For the response y_1, the estimated variance components are $\hat{\sigma}_\varepsilon^2 = 0.09695$ and $\hat{\sigma}_\gamma^2 = 0.9703$, yielding an estimated degree of correlation of $\hat{\eta} = 10.01$. For the response y_2, the estimated variance components amount to $\hat{\sigma}_\varepsilon^2 = 0.1003$ and $\hat{\sigma}_\gamma^2 = 0.1408$, yielding an estimated degree of correlation of $\hat{\eta} = 1.4038$.

From the tables, it can be seen that the significant effects are different for the two responses under study. For example, the linear effect of the first factor x_1 turns out to be significant for the model in y_1, whereas x_1 has no significant explaining value for y_2.

Table 3.4: Coefficient estimates, standard errors, t-statistics and p-values for response y_2 in the pastry dough mixing experiment.

Term	Estimate	St. error	t-value	p-value
Intercept	77.0930	0.1854	415.88	$<.0001$
x1	0.0689	0.0746	0.92	0.3742
x2	-0.2844	0.0746	-3.81	0.0025
x3	0.2294	0.0746	3.07	0.0096
x1x2	0.0582	0.0878	0.66	0.5201
x1x3	0.1168	0.0878	1.33	0.2082
x2x3	-0.3648	0.0878	-4.15	0.0013
x11	0.1201	0.1905	0.63	0.5402
x22	-0.1799	0.1905	-0.94	0.3638
x33	0.4551	0.1905	2.39	0.0342

3.3.2   The analysis of a split-plot experiment

Consider as an illustration the split-plot experiment described in Kowalski, Cornell and Vining (2002). The experiment has 28 runs and is a modified version of an example in Cornell (1990). It involves the production of vinyl for automobile seat covers. In the experiment, the effect of five factors on the thickness of the vinyl are investigated. Three of the factors are mixture components and two of them are so-called process variables. As in ordinary mixture experiments (see Section 1.4.4), the component proportions sum to one. In this example, the response of interest does not only depend on these proportions, but also on the effects of the process variables. The mixture components in the experiment are three plasticizers whose proportions are represented by x_1, x_2 and x_3. The two process variables studied are rate of extrusion (z_1) and temperature of drying (z_2).

The factor level combinations used in the experiment are graphically displayed in Figure 3.4. Two levels were used for each process variable. For the mixture components, the points of the second order lattice design in Figure 1.1 were used, as well as the centroid of the simplex. The experiment was carried out by randomly selecting a combination of the levels of the process variables and running all blends at this combination. Next, another combination of the process variable levels is randomly chosen and all blends are run at this combination. This procedure was repeated until all combinations of the process variables had been performed. The experiment was therefore conducted in a split-plot format. The process variables are the whole plot factors of the experiment, whereas the mixture components are the sub-plot factors. The data are shown in Table 3.5.

A main effects plus two factor interactions model was assumed for the process variables z_1 and z_2. For the mixture components, the quadratic

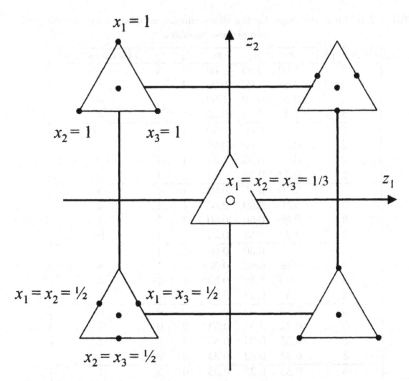

$x_1 = 1$

z_2

$x_2 = 1$ $x_3 = 1$

$x_1 = x_2 = x_3 = 1/3$ z_1

$x_1 = x_2 = \frac{1}{2}$ $x_1 = x_3 = \frac{1}{2}$

$x_2 = x_3 = \frac{1}{2}$

Figure 3.4: Experimental design for the three-component mixture experiment with two process variables. • is a design point, ○ is a design point replicated twelve times.

mixture model (1.41) was used. The main effects of the process variables were crossed with the linear blending terms only, so that the assumed model was

$$y_{ij} = \sum_{i=1}^{3} \beta_i x_i + \sum_{i=1}^{m-1} \sum_{j=i+1}^{m} \beta_{ij} x_i x_j + \alpha z_1 z_2 + \sum_{i=1}^{3} \sum_{j=1}^{2} \delta_{ij} x_i z_j + \gamma_i + \varepsilon_{ij}.$$

We have computed the REML estimates of all factor effects and the variance components σ_ε^2 and σ_γ^2 using PROC MIXED in SAS. The following commands were used:

```
data kcv;
infile 'vinyl.dat';
input wp x1 x2 x3 z1 z2 y;
x1x2=x1*x2;
x1x3=x1*x3;
x2x3=x2*x3;
```

Table 3.5: Vinyl thickness for the three-component mixture experiment with two process variables.

Whole plot	x_1	x_2	x_3	z_1	z_2	Thickness value
1	1.00	0.00	0.00	-1	1	10
1	0.00	1.00	0.00	-1	1	8
1	0.00	0.00	1.00	-1	1	3
1	0.33	0.33	0.33	-1	1	8
2	1.00	0.00	0.00	1	-1	10
2	0.00	1.00	0.00	1	-1	5
2	0.00	0.00	1.00	1	-1	9
2	0.33	0.33	0.33	1	-1	9
3	0.50	0.50	0.00	1	1	5
3	0.50	0.00	0.50	1	1	4
3	0.00	0.50	0.50	1	1	7
3	0.33	0.33	0.33	1	1	10
4	0.50	0.50	0.00	-1	-1	7
4	0.50	0.00	0.50	-1	-1	8
4	0.00	0.50	0.50	-1	-1	4
4	0.33	0.33	0.33	-1	-1	7
5	0.33	0.33	0.33	0	0	8
5	0.33	0.33	0.33	0	0	7
5	0.33	0.33	0.33	0	0	7
5	0.33	0.33	0.33	0	0	8
6	0.33	0.33	0.33	0	0	7
6	0.33	0.33	0.33	0	0	8
6	0.33	0.33	0.33	0	0	9
6	0.33	0.33	0.33	0	0	9
7	0.33	0.33	0.33	0	0	12
7	0.33	0.33	0.33	0	0	10
7	0.33	0.33	0.33	0	0	9
7	0.33	0.33	0.33	0	0	11

```
z1z2=z1*z2;
x1z1=x1*z1;
x2z1=x2*z1;
x3z1=x3*z1;
x1z2=x1*z2;
x2z2=x2*z2;
x3z2=x3*z2;
proc mixed;
class wp;
model y = z1z2 x1 x2 x3 x1x2 x1x3 x2x3 x1z1 x2z1
          x3z1 x1z2 x2z2 x3z2 / solution noint;
```

Table 3.6: Coefficient estimates, standard errors, t-statistics and p-values for the vinyl thickness example.

Term	Estimate	St. error	t-value	p-value
$z_1 z_2$	-1.1945	0.7748	-1.54	0.1575
$x1$	8.8940	1.2189	7.30	< .0001
$x2$	5.3940	1.2189	4.43	0.0017
$x3$	4.8940	1.2189	4.02	0.0030
$x_1 x_2$	3.7413	4.7192	0.79	0.4483
$x_1 x_3$	4.7413	4.7192	1.00	0.3413
$x_2 x_3$	9.7413	4.7192	2.06	0.0690
$x_1 z_1$	-2.0074	1.1543	-1.74	0.1160
$x_2 z_1$	0.7426	1.1543	0.64	0.5361
$x_3 z_1$	1.9926	1.1543	1.73	0.1184
$x_1 z_2$	-2.0086	1.1543	-1.74	0.1158
$x_2 z_2$	2.2414	1.1543	1.94	0.0841
$x_3 z_2$	-1.0086	1.1543	-0.87	0.4049

```
random wp;
run;
```

The option NOINT is specified because no intercept is included in the statistical model under investigation. The two variance components were estimated as $\sigma_\varepsilon^2 = 1.9470$ and $\sigma_\gamma^2 = 1.5876$, yielding an estimated degree of correlation of $\hat{\eta} = 0.8154$. The factor effect estimates are displayed in Table 3.6, along with the estimated standard errors and the p-values. The p-values computed by SAS were obtained from a t-distribution with nine degrees of freedom. Apart from the linear blending effects, none of the factor effects appears to be significantly different from zero at the $\alpha = 0.05$ level. It should be pointed out, however, that the linear blending effects are not direct measures of the component effects as they cannot be interpreted independently of each other.

3.4   Information matrix

Using Theorem 18.2.8 of Harville (1997), we derive from (3.8) that

$$\mathbf{V}_i^{-1} = \frac{1}{\sigma_\varepsilon^2}(\mathbf{I}_{k_i \times k_i} - \frac{\eta}{1 + k_i\eta}\mathbf{1}_{k_i}\mathbf{1}_{k_i}'). \tag{3.14}$$

Since $\mathbf{V}$ is block diagonal, the information matrices of the experiments described in this book can be written as in (2.8). Substituting (3.14) in (2.8) yields

$$
\begin{aligned}
\mathbf{M} &= \frac{1}{\sigma_\varepsilon^2} \sum_{i=1}^{b} \mathbf{X}_i'(\mathbf{I}_{k_i \times k_i} - \frac{\eta}{1 + k_i \eta} \mathbf{1}_{k_i} \mathbf{1}_{k_i}') \mathbf{X}_i, \\
&= \frac{1}{\sigma_\varepsilon^2} \sum_{i=1}^{b} (\mathbf{X}_i' - \frac{\eta}{1 + k_i \eta} \mathbf{X}_i' \mathbf{1}_{k_i} \mathbf{1}_{k_i}') \mathbf{X}_i, \\
&= \frac{1}{\sigma_\varepsilon^2} (\sum_{i=1}^{b} \mathbf{X}_i' \mathbf{X}_i - \frac{\eta}{1 + k_i \eta} \mathbf{X}_i' \mathbf{1}_{k_i} \mathbf{1}_{k_i}' \mathbf{X}_i), \\
&= \frac{1}{\sigma_\varepsilon^2} \Big\{ \mathbf{X}' \mathbf{X} - \sum_{i=1}^{b} \frac{\eta}{1 + k_i \eta} (\mathbf{X}_i' \mathbf{1}_{k_i})(\mathbf{X}_i' \mathbf{1}_{k_i})' \Big\},
\end{aligned}
\tag{3.15}
$$

where $\mathbf{X}_i$ is the part of $\mathbf{X}$ corresponding to the ith group. Using (1.55), this information matrix can be written as a sum of outer products of vectors:

$$
\mathbf{M} = \frac{1}{\sigma_\varepsilon^2} \Big\{ \sum_{i=1}^{b} \sum_{j=1}^{k_i} \mathbf{f}(\mathbf{x}_{ij}) \mathbf{f}'(\mathbf{x}_{ij}) - \sum_{i=1}^{b} \frac{\eta}{1 + k_i \eta} (\mathbf{X}_i' \mathbf{1}_{k_i})(\mathbf{X}_i' \mathbf{1}_{k_i})' \Big\}, \tag{3.16}
$$

where $\mathbf{f}(\mathbf{x}_{ij})$ is the polynomial expansion corresponding to the jth observation within the ith group, or equivalently, the jth row of $\mathbf{X}_i$. In comparison with expression (2.9), expression (3.16) has the advantage of interpretability. Firstly, an outer product $\mathbf{f}(\mathbf{x}_{ij}) \mathbf{f}'(\mathbf{x}_{ij})$ is included for each design point of the experiment. Secondly, a fraction $\eta/(1 + k_i \eta)$ of the outer product $(\mathbf{X}_i' \mathbf{1}_{k_i})(\mathbf{X}_i' \mathbf{1}_{k_i})'$ is subtracted for each group in the experiment. Each $\mathbf{X}_i' \mathbf{1}_{k_i}$ is equal to

$$
\mathbf{X}_i' \mathbf{1}_{k_i} = \sum_{j=1}^{k_i} \mathbf{f}(\mathbf{x}_{ij}),
$$

that is the p-dimensional vector containing the sums of the individual rows of $\mathbf{X}_i$. In addition, (3.16) allows a fast update of the information matrix after a design change. For most design problems, a faster update is possible than by using (2.9).

In the sequel of this section, it is shown how to update the information matrix after a design change. For each modification considered, the update is accomplished by adding and/or subtracting one or more vector outer products. The determinant and the inverse of the information matrix can be updated by repeatedly applying the update formulae given in Sections 1.7 and 2.1. Assume that the current information matrix, the current design matrix, the part of the current design matrix corresponding to the ith group and the current variance-covariance matrix are denoted by $\mathbf{M}$, $\mathbf{X}$, $\mathbf{X}_i$ and $\mathbf{V}$ respectively. The corresponding modified matrices will be denoted by $\mathbf{M}^*$,

$\mathbf{X}^*$, $\mathbf{X}_i^*$ and $\mathbf{V}^*$ respectively. For notational simplicity, assume without loss of generality that $\sigma_\varepsilon^2 = 1$.

3.4.1   Addition of a design point

Adding a design point $\mathbf{a}$ to the ith group will affect both $\mathbf{X}$ and $\mathbf{V}$. An extra row $\mathbf{f}'(\mathbf{a})$ will be added to the design matrix $\mathbf{X}$, more specifically to that part of the design matrix corresponding to the ith group, $\mathbf{X}_i$. The dimension of $\mathbf{V}_i$ and thus of $\mathbf{V}$ will be increased by one. The updated information matrix $\mathbf{M}^*$ is then given by

$$
\begin{aligned}
\mathbf{X}^{*\prime}\mathbf{V}^{*-1}\mathbf{X}^* = \mathbf{X}'\mathbf{V}^{-1}\mathbf{X} &+ \mathbf{f}(\mathbf{a})\mathbf{f}'(\mathbf{a}) \\
&+ \frac{\eta}{1+k_i\eta}(\mathbf{X}_i'\mathbf{1}_{k_i})(\mathbf{X}_i'\mathbf{1}_{k_i})' \\
&- \frac{\eta}{1+(k_i+1)\eta}(\mathbf{X}_i^{*\prime}\mathbf{1}_{(k_i+1)})(\mathbf{X}_i^{*\prime}\mathbf{1}_{(k_i+1)})',
\end{aligned}
\tag{3.17}
$$

where

$$
\mathbf{X}_i^{*\prime}\mathbf{1}_{(k_i+1)} = \mathbf{X}_i'\mathbf{1}_{k_i} + \mathbf{f}(\mathbf{a}).
\tag{3.18}
$$

When the first observation of a new group is added to the design, (3.17) can be simplified. In that case, $k_{b+1} = 0$, $\mathbf{X}_{b+1} = \mathbf{0}$ and $\mathbf{X}_{b+1}^* = \mathbf{f}'(\mathbf{a})$, so that

$$
\mathbf{M}^* = \mathbf{X}'\mathbf{V}^{-1}\mathbf{X} + \frac{1}{1+\eta}\mathbf{f}(\mathbf{a})\mathbf{f}'(\mathbf{a}).
\tag{3.19}
$$

3.4.2   Deletion of a design point

Deleting a design point $\mathbf{a}$ from the ith group will also affect $\mathbf{X}$ and $\mathbf{V}$. One row $\mathbf{f}'(\mathbf{a})$ will be removed from the design matrix $\mathbf{X}$, more specifically from that part of the design matrix corresponding to the ith group, $\mathbf{X}_i$. The dimension of $\mathbf{V}_i$ and thus of $\mathbf{V}$ will be decreased by one. The updated information matrix $\mathbf{M}^*$ is then given by

$$
\begin{aligned}
\mathbf{X}^{*\prime}\mathbf{V}^{*-1}\mathbf{X}^* = \mathbf{X}'\mathbf{V}^{-1}\mathbf{X} &- \mathbf{f}(\mathbf{a})\mathbf{f}'(\mathbf{a}) \\
&+ \frac{\eta}{1+k_i\eta}(\mathbf{X}_i'\mathbf{1}_{k_i})(\mathbf{X}_i'\mathbf{1}_{k_i})' \\
&- \frac{\eta}{1+(k_i-1)\eta}(\mathbf{X}_i^{*\prime}\mathbf{1}_{(k_i-1)})(\mathbf{X}_i^{*\prime}\mathbf{1}_{(k_i-1)})',
\end{aligned}
\tag{3.20}
$$

where

$$
\mathbf{X}_i^{*\prime}\mathbf{1}_{(k_i-1)} = \mathbf{X}_i'\mathbf{1}_{k_i} - \mathbf{f}(\mathbf{a}).
\tag{3.21}
$$

When the only observation of the ith group is removed from the design, (3.20) can be simplified. In that case, $k_i = 1$, $\mathbf{X}_i = \mathbf{f}'(\mathbf{a})$ and $\mathbf{X}_i^* = \mathbf{0}$, so that

$$M^* = X'V^{-1}X - \frac{1}{1+\eta}f(a)f'(a). \tag{3.22}$$

3.4.3   Substitution of a design point

Substituting a design point a by a point b in the ith group affects X, but not V. The row $f'(a)$ will be replaced by $f'(b)$ in that part of the design matrix corresponding to the ith group, X_i. The updated information matrix M^* is then given by

$$\begin{aligned}
X^{*\prime}V^{-1}X^* = {}& X'V^{-1}X - f(a)f'(a) + f(b)f'(b) \\
& + \frac{\eta}{1+k_i\eta}(X_i'1_{k_i})(X_i'1_{k_i})' \\
& - \frac{\eta}{1+k_i\eta}(X_i^{*\prime}1_{k_i})(X_i^{*\prime}1_{k_i})',
\end{aligned} \tag{3.23}$$

where

$$X_i^{*\prime}1_{(k_i)} = X_i'1_{k_i} - f(a) + f(b). \tag{3.24}$$

3.4.4   Interchange of two design points from different groups

Moving a design point a from group i to group j and a design point b from group j to group i affects X_i and X_j in the following fashion:

$$X_i^{*\prime}1_{k_i} = X_i'1_{k_i} - f(a) + f(b),$$

and

$$X_j^{*\prime}1_{k_j} = X_j'1_{k_j} + f(a) - f(b).$$

The updated information matrix M^* is then given by

$$\begin{aligned}
X^{*\prime}V^{-1}X^* = {}& X'V^{-1}X \\
& + \frac{\eta}{1+k_i\eta}(X_i'1_{k_i})(X_i'1_{k_i})' \\
& + \frac{\eta}{1+k_j\eta}(X_j'1_{k_j})(X_j'1_{k_j})' \\
& - \frac{\eta}{1+k_i\eta}(X_i^{*\prime}1_{k_i})(X_i^{*\prime}1_{k_i})' \\
& - \frac{\eta}{1+k_j\eta}(X_j^{*\prime}1_{k_j})(X_j^{*\prime}1_{k_j})'.
\end{aligned} \tag{3.25}$$

3.5   Equivalence of OLS and GLS

In some cases, ordinary and generalized least squares prove to be equivalent, i.e.

$$(\mathbf{X}'\mathbf{V}^{-1}\mathbf{X})^{-1}\mathbf{X}'\mathbf{V}^{-1}\mathbf{y} = (\mathbf{X}'\mathbf{X})^{-1}\mathbf{X}'\mathbf{y},$$

implying that, in these cases, error variance knowledge is no longer necessary for model estimation purposes. It should, however, be emphasized that knowledge of both variance components remains essential for statistical inference regarding the estimated model.

An important consequence of the equivalence of OLS and GLS is that the variance-covariance matrix (3.10) is not only valid asymptotically. This result will be used in a simulation study in Section 3.6.

3.5.1   Saturated designs

A saturated design is a design in which the number of observations n is equal to the number of model parameters p. In that case, the design matrix $\mathbf{X}$ is a square matrix. Provided $\mathbf{X}$ is regular, the GLS estimator then becomes

$$
\begin{aligned}
(\mathbf{X}'\mathbf{V}^{-1}\mathbf{X})^{-1}\mathbf{X}'\mathbf{V}^{-1}\mathbf{y} &= \mathbf{X}^{-1}\mathbf{V}(\mathbf{X}')^{-1}\mathbf{X}'\mathbf{V}^{-1}\mathbf{y}, \\
&= \mathbf{X}^{-1}\mathbf{V}\mathbf{V}^{-1}\mathbf{y}, \\
&= \mathbf{X}^{-1}\mathbf{y},
\end{aligned}
\tag{3.26}
$$

and the OLS estimator is equal to

$$
\begin{aligned}
(\mathbf{X}'\mathbf{X})^{-1}\mathbf{X}'\mathbf{y} &= \mathbf{X}^{-1}(\mathbf{X}')^{-1}\mathbf{X}'\mathbf{y}, \\
&= \mathbf{X}^{-1}\mathbf{y}.
\end{aligned}
\tag{3.27}
$$

As a result, OLS and GLS are equivalent.

3.5.2   Orthogonally blocked designs with homogeneous block sizes

The concept of orthogonal blocking was introduced in Section 2.3.4 for the fixed block effects model. Khuri (1992) shows that, for an orthogonally blocked design, OLS and GLS produce the same estimate for $\tilde{\beta}$, that is for all elements of β apart from the intercept. He also shows that the same estimate is obtained from the intra-block analysis (see Section 2.3.3) as well. The proofs of these equivalencies are valid for both homogeneous and heterogeneous block sizes.

For homogeneous block sizes $k = k_1 = k_2 = \cdots = k_b$, we can show that the OLS and the GLS estimator for the intercept are equivalent as well. In addition, these estimators are identical to the intra-block estimator for the intercept obtained from (2.27) and (2.29). Let us denote by $[\; \hat{\beta}_{0,\text{OLS}} \; \hat{\tilde{\beta}}_{\text{OLS}} \;]$ the OLS estimator for $\beta = [\; \beta_0 \; \tilde{\beta}' \;]'$ and by $[\; \hat{\beta}_{0,\text{GLS}} \; \hat{\tilde{\beta}}_{\text{GLS}} \;]$ the GLS estimator. Further, assume without loss of generality that $\sigma_\varepsilon^2 = 1$ and denote $\mathbf{X} = [\; \mathbf{1}_n \; \tilde{\mathbf{X}} \;]$. Using Theorem 8.5.11 of Harville (1997), the OLS estimator (1.8) can be written as

$$
\begin{bmatrix} \hat{\beta}_{0,\text{OLS}} \\ \hat{\tilde{\beta}}_{\text{OLS}} \end{bmatrix} = \begin{bmatrix} \mathbf{1}_n'\mathbf{1}_n & \mathbf{1}_n'\tilde{\mathbf{X}} \\ \tilde{\mathbf{X}}'\mathbf{1}_n & \tilde{\mathbf{X}}'\tilde{\mathbf{X}} \end{bmatrix}^{-1} \begin{bmatrix} \mathbf{1}_n'\mathbf{y} \\ \tilde{\mathbf{X}}'\mathbf{y} \end{bmatrix},
$$

$$
= \begin{bmatrix} (\mathbf{1}_n'\mathbf{1}_n)^{-1} + (\mathbf{1}_n'\mathbf{1}_n)^{-1}\mathbf{1}_n'\tilde{\mathbf{X}}\mathbf{P}_1 & -(\mathbf{1}_n'\mathbf{1}_n)^{-1}\mathbf{1}_n'\tilde{\mathbf{X}}\mathbf{P}_2 \\ -\mathbf{P}_1 & \mathbf{P}_2 \end{bmatrix} \begin{bmatrix} \mathbf{1}_n'\mathbf{y} \\ \tilde{\mathbf{X}}'\mathbf{y} \end{bmatrix},
$$

$$
= \begin{bmatrix} (\mathbf{1}_n'\mathbf{1}_n)^{-1}\mathbf{1}_n'\mathbf{y} + (\mathbf{1}_n'\mathbf{1}_n)^{-1}\mathbf{1}_n'\tilde{\mathbf{X}}(\mathbf{P}_1\mathbf{1}_n'\mathbf{y} - \mathbf{P}_2\tilde{\mathbf{X}}'\mathbf{y}) \\ -\mathbf{P}_1\mathbf{1}_n'\mathbf{y} + \mathbf{P}_2\tilde{\mathbf{X}}'\mathbf{y} \end{bmatrix},
$$

where

$$
\mathbf{P}_1 = \mathbf{P}_2\tilde{\mathbf{X}}'\mathbf{1}_n(\mathbf{1}_n'\mathbf{1}_n)^{-1},
$$

and

$$
\mathbf{P}_2 = \{\tilde{\mathbf{X}}'\tilde{\mathbf{X}} - \tilde{\mathbf{X}}'\mathbf{1}_n(\mathbf{1}_n'\mathbf{1}_n)^{-1}\mathbf{1}_n'\tilde{\mathbf{X}}\}^{-1}.
$$

As a result,

$$
\hat{\beta}_{0,\text{OLS}} = (\mathbf{1}_n'\mathbf{1}_n)^{-1}\mathbf{1}_n'\mathbf{y} - (\mathbf{1}_n'\mathbf{1}_n)^{-1}\mathbf{1}_n'\tilde{\mathbf{X}}\hat{\tilde{\beta}}_{\text{OLS}},
$$

or

$$
\hat{\beta}_{0,\text{OLS}} = \frac{1}{n}\mathbf{1}_n'\mathbf{y} - \frac{1}{n}\mathbf{1}_n'\tilde{\mathbf{X}}\hat{\tilde{\beta}}_{\text{OLS}}, \tag{3.28}
$$

since $\mathbf{1}_n'\mathbf{1}_n = n$. The GLS estimator (3.9) is given by

$$
\begin{bmatrix} \hat{\beta}_{0,\text{GLS}} \\ \hat{\tilde{\beta}}_{\text{GLS}} \end{bmatrix} = \begin{bmatrix} \mathbf{1}_n'\mathbf{V}^{-1}\mathbf{1}_n & \mathbf{1}_n'\mathbf{V}^{-1}\tilde{\mathbf{X}} \\ \tilde{\mathbf{X}}'\mathbf{V}^{-1}\mathbf{1}_n & \tilde{\mathbf{X}}'\mathbf{V}^{-1}\tilde{\mathbf{X}} \end{bmatrix}^{-1} \begin{bmatrix} \mathbf{1}_n'\mathbf{V}^{-1}\mathbf{y} \\ \tilde{\mathbf{X}}'\mathbf{V}^{-1}\mathbf{y} \end{bmatrix}.
$$

Denoting $\tilde{\mathbf{V}} = \mathbf{V}_1 = \mathbf{V}_2 = \cdots = \mathbf{V}_b$, we have from Harville's (1997) Theorem 18.2.8 that

$$
\tilde{\mathbf{V}}^{-1} = \sigma_\varepsilon^{-2}(\mathbf{I}_{k\times k} - \frac{\eta}{1+k\eta}\mathbf{1}_k\mathbf{1}_k').
$$

Hence

$$
\mathbf{1}_n'\mathbf{V}^{-1}\mathbf{1}_n = b\mathbf{1}_k'\tilde{\mathbf{V}}^{-1}\mathbf{1}_k,
$$

$$
= \sigma_\varepsilon^{-2}b\mathbf{1}_k'(\mathbf{I}_{k\times k} - \frac{\eta}{1+k\eta}\mathbf{1}_k\mathbf{1}_k')\mathbf{1}_k,
$$

$$
= \sigma_\varepsilon^{-2}b(k - \frac{\eta}{1+k\eta}k^2),
$$

$$
= \sigma_\varepsilon^{-2}n(\frac{1}{1+k\eta}),
$$

$$
= nc,
$$

where $c = \sigma_\epsilon^{-2}(1 + k\eta)^{-1}$. The GLS estimator then becomes

$$
\begin{bmatrix} \hat{\beta}_{0,\text{GLS}} \\ \hat{\beta}_{\text{GLS}} \end{bmatrix} = \begin{bmatrix} nc & \mathbf{1}_n' \mathbf{V}^{-1} \tilde{\mathbf{X}} \\ \tilde{\mathbf{X}}' \mathbf{V}^{-1} \mathbf{1}_n & \tilde{\mathbf{X}}' \mathbf{V}^{-1} \tilde{\mathbf{X}} \end{bmatrix}^{-1} \begin{bmatrix} \mathbf{1}_n' \mathbf{V}^{-1} \mathbf{y} \\ \tilde{\mathbf{X}}' \mathbf{V}^{-1} \mathbf{y} \end{bmatrix},
$$

$$
= \begin{bmatrix} (nc)^{-1} + (nc)^{-1} \mathbf{1}_n' \mathbf{V}^{-1} \tilde{\mathbf{X}} \mathbf{Q}_1 & -(nc)^{-1} \mathbf{1}_n' \mathbf{V}^{-1} \tilde{\mathbf{X}} \mathbf{Q}_2 \\ -\mathbf{Q}_1 & \mathbf{Q}_2 \end{bmatrix} \begin{bmatrix} \mathbf{1}_n' \mathbf{V}^{-1} \mathbf{y} \\ \tilde{\mathbf{X}}' \mathbf{V}^{-1} \mathbf{y} \end{bmatrix},
$$

$$
= \begin{bmatrix} (nc)^{-1} \mathbf{1}_n' \mathbf{y} + (nc)^{-1} \mathbf{1}_n' \mathbf{V}^{-1} \tilde{\mathbf{X}} (\mathbf{Q}_1 \mathbf{1}_n' \mathbf{V}^{-1} \mathbf{y} - \mathbf{Q}_2 \tilde{\mathbf{X}}' \mathbf{V}^{-1} \mathbf{y}) \\ -\mathbf{Q}_1 \mathbf{1}_n' \mathbf{V}^{-1} \mathbf{y} + \mathbf{Q}_2 \tilde{\mathbf{X}}' \mathbf{V}^{-1} \mathbf{y} \end{bmatrix},
$$

where

$$
\mathbf{Q}_1 = \mathbf{Q}_2 \tilde{\mathbf{X}}' \mathbf{V}^{-1} \mathbf{1}_n (nc)^{-1},
$$

and

$$
\mathbf{Q}_2 = \{\tilde{\mathbf{X}}' \mathbf{V}^{-1} \tilde{\mathbf{X}} - \tilde{\mathbf{X}}' \mathbf{V}^{-1} \mathbf{1}_n (nc)^{-1} \mathbf{1}_n' \mathbf{V}^{-1} \tilde{\mathbf{X}}\}^{-1}.
$$

As a result,

$$
\hat{\beta}_{0,\text{GLS}} = (nc)^{-1} \mathbf{1}_n' \mathbf{V}^{-1} \mathbf{y} - (nc)^{-1} \mathbf{1}_n' \mathbf{V}^{-1} \tilde{\mathbf{X}} \hat{\beta}_{\text{GLS}}.
$$

Since

$$
\mathbf{1}_n' \mathbf{V}^{-1} \mathbf{y} = \sum_{i=1}^{b} \mathbf{1}_k' \tilde{\mathbf{V}}^{-1} \mathbf{y}_i,
$$

$$
= \sigma_\epsilon^{-2} \sum_{i=1}^{b} \mathbf{1}_k' (\mathbf{I}_{k \times k} - \frac{\eta}{1 + k\eta} \mathbf{1}_k \mathbf{1}_k') \mathbf{y}_i,
$$

$$
= \sigma_\epsilon^{-2} \sum_{i=1}^{b} (\mathbf{1}_k' \mathbf{y}_i - \frac{k\eta}{1 + k\eta} \mathbf{1}_k' \mathbf{y}_i),
$$

$$
= \sigma_\epsilon^{-2} (1 - \frac{k\eta}{1 + k\eta}) \sum_{i=1}^{b} \mathbf{1}_k' \mathbf{y}_i,
$$

$$
= \sigma_\epsilon^{-2} (1 - \frac{k\eta}{1 + k\eta}) \mathbf{1}_n' \mathbf{y},
$$

$$
= c \mathbf{1}_n' \mathbf{y},
$$

where $\mathbf{y}_i$ is that part of $\mathbf{y}$ corresponding to the ith block, and

$$
\begin{aligned}
\mathbf{1}'_n \mathbf{V}^{-1} \tilde{\mathbf{X}} &= \sum_{i=1}^{b} \mathbf{1}'_k \tilde{\mathbf{V}}^{-1} \tilde{\mathbf{X}}_i, \\
&= \sigma_\varepsilon^{-2} \sum_{i=1}^{b} \mathbf{1}'_k (\mathbf{I}_{k \times k} - \frac{\eta}{1+k\eta} \mathbf{1}_k \mathbf{1}'_k) \tilde{\mathbf{X}}_i, \\
&= \sigma_\varepsilon^{-2} \sum_{i=1}^{b} (\mathbf{1}'_k \tilde{\mathbf{X}}_i - \frac{k\eta}{1+k\eta} \mathbf{1}'_k \tilde{\mathbf{X}}_i), \\
&= \sigma_\varepsilon^{-2} (1 - \frac{k\eta}{1+k\eta}) \sum_{i=1}^{b} \mathbf{1}'_k \tilde{\mathbf{X}}_i, \\
&= \sigma_\varepsilon^{-2} (1 - \frac{k\eta}{1+k\eta}) \sum_{i=1}^{b} \frac{k}{n} \mathbf{1}'_n \tilde{\mathbf{X}}, \\
&= c \frac{bk}{n} \mathbf{1}'_n \tilde{\mathbf{X}}, \\
&= c \mathbf{1}'_n \tilde{\mathbf{X}},
\end{aligned}
$$

this estimator becomes

$$
\begin{aligned}
\hat{\beta}_{0,\mathrm{GLS}} &= (nc)^{-1} c \mathbf{1}'_n \mathbf{y} - (nc)^{-1} c \mathbf{1}'_n \tilde{\mathbf{X}} \hat{\hat{\beta}}_{\mathrm{GLS}}, \\
&= \frac{1}{n} \mathbf{1}'_n \mathbf{y} - \frac{1}{n} \mathbf{1}'_n \tilde{\mathbf{X}} \hat{\hat{\beta}}_{\mathrm{GLS}}.
\end{aligned}
\tag{3.29}
$$

Since $\hat{\hat{\beta}}_{\mathrm{OLS}} = \hat{\hat{\beta}}_{\mathrm{GLS}}$ for an orthogonally blocked design, (3.28) and (3.29) are identical. As a consequence, the OLS and the GLS estimator for β are equivalent for orthogonally blocked designs with homogeneous block sizes. An illustration of an orthogonally blocked experiment with homogeneous block sizes is given in Section 2.3.5. It can be verified that ordinary least squares estimation, generalized least squares estimation and intra-block analysis all produce the estimated model given in Tables 2.3 and 2.4.

When the block sizes are heterogeneous, the estimate for the intercept depends on the estimation method used. Heterogeneous block sizes frequently occur when an orthogonally blocked second order standard design is used. For example, Box and Hunter (1957) propose an orthogonally blocked central composite design for three variables with 2 factorial blocks of size 6 and one axial block of size 8. Another example of an orthogonally blocked experiment with heterogeneous block sizes is given in Table 3.7. This 54-run experiment was carried out in the laboratory of a multinational producing ingredients for food and beverage applications to investigate the effect of adding salt on the yield (expressed in %) of a starch extraction process. Five blocks corresponding to batches of raw material originating from five of the countries in which the company was active were used in the experi-

Table 3.7: Orthogonally blocked starch extraction experiment.

x	Block 1			Block 2			Block 3	Block 4	Block 5
-1	40.5	39.4	42.3	40.8	39.4	38.6	49.3	48.5	38.1
-0.5	43.2	41.7	41.3	40.1	43.5	44.4	51.0	52.8	44.7
0	50.4	43.6	46.6	45.7	42.8	48.4	55.8	62.2	50.6
0	47.4	45.0	48.5	47.6	48.9	45.8	59.4	55.5	52.1
0.5	49.1	53.1	51.3	46.9	47.6	50.8	56.4	60.5	51.3
1	50.0	48.3	50.1	49.1	48.3	50.2	60.0	60.1	52.6

ment. Two of the blocks contain 18 runs, whereas the others contains only six runs. Five equally spaced salt levels were used in the experiment and the middle level was replicated twice as much as the other levels. As the yield was expected to increase at a slackening pace, a quadratic model was fitted. GLS estimation gave

$$E(y) = 51.41 + 5.51x - 2.58x^2, \qquad (3.30)$$

whereas OLS produced

$$E(y) = 49.44 + 5.51x - 2.58x^2, \qquad (3.31)$$

and the intra-block analysis yielded

$$E(y) = 51.45 + 5.51x - 2.58x^2 \qquad (3.32)$$

when (2.27) was used, and

$$E(y) = 49.44 + 5.51x - 2.58x^2 \qquad (3.33)$$

when (2.29) was used. The first model was obtained using the ratio of the REML estimates $\hat{\sigma}_\gamma^2 = 26.68$ and $\hat{\sigma}_\varepsilon^2 = 4.21$ as an estimate for η. It is clear that, apart from the intercept, all models are identical. It can also be verified that the intercepts in the models (3.31) and (3.33) are identical. This is generally true for orthogonally blocked experiments.

If the variance components σ_γ^2 and σ_ε^2 are known, the GLS estimator is the most efficient one. The intra-block estimator (2.27) of the intercept is slightly less efficient. If the variance components are unknown, as is usually the case in practice, the intra-block estimator (2.27) sometimes becomes more efficient than the feasible GLS estimator. In cases where the interest is in estimating the location of a stationary point on a response surface or in estimating the difference between the responses for two combinations of the factor levels, the value of the intercept does not matter. However, the goals of an experiment often include the prediction of the response for certain combinations of the factor levels. This is important when a target value for the response has to be achieved.

3.5.3 Crossed split-plot designs

Crossed split-plot designs differ from non-crossed split-plot designs in that every combination of levels of the sub-plot variables appears in each whole plot. Each whole plot then has an equal number of sub-plots, as well as equal levels of the sub-plot variables. In non-crossed split-plot designs, each whole plot may have a different number of sub-plots and the levels of the sub-plot variables need no longer be identical across whole plots. In general, the design matrices $\mathbf{X} = [\ \mathbf{X}_1' \ \mathbf{X}_2' \ \ldots \ \mathbf{X}_b' \ ]'$ of crossed and non-crossed split-plot designs can be written as

$$
\begin{bmatrix}
\mathbf{f}'(\mathbf{w}_1, \mathbf{s}_1) \\
\mathbf{f}'(\mathbf{w}_1, \mathbf{s}_2) \\
\vdots \\
\mathbf{f}'(\mathbf{w}_1, \mathbf{s}_k) \\
\cdots\cdots \\
\mathbf{f}'(\mathbf{w}_2, \mathbf{s}_1) \\
\mathbf{f}'(\mathbf{w}_2, \mathbf{s}_2) \\
\vdots \\
\mathbf{f}'(\mathbf{w}_2, \mathbf{s}_k) \\
\cdots\cdots \\
\vdots \\
\cdots\cdots \\
\mathbf{f}'(\mathbf{w}_b, \mathbf{s}_1) \\
\mathbf{f}'(\mathbf{w}_b, \mathbf{s}_2) \\
\vdots \\
\mathbf{f}'(\mathbf{w}_b, \mathbf{s}_k)
\end{bmatrix}
\quad \text{and} \quad
\begin{bmatrix}
\mathbf{f}'(\mathbf{w}_1, \mathbf{s}_{11}) \\
\mathbf{f}'(\mathbf{w}_1, \mathbf{s}_{12}) \\
\vdots \\
\mathbf{f}'(\mathbf{w}_1, \mathbf{s}_{1k_1}) \\
\cdots\cdots \\
\mathbf{f}'(\mathbf{w}_2, \mathbf{s}_{21}) \\
\mathbf{f}'(\mathbf{w}_2, \mathbf{s}_{22}) \\
\vdots \\
\mathbf{f}'(\mathbf{w}_2, \mathbf{s}_{2k_2}) \\
\cdots\cdots \\
\vdots \\
\cdots\cdots \\
\mathbf{f}'(\mathbf{w}_b, \mathbf{s}_{b1}) \\
\mathbf{f}'(\mathbf{w}_b, \mathbf{s}_{b2}) \\
\vdots \\
\mathbf{f}'(\mathbf{w}_b, \mathbf{s}_{bk_w})
\end{bmatrix}
$$

respectively, where the whole plot factors are denoted by $\mathbf{w}$ and the sub-plot factors by $\mathbf{s}$. Designs that fall within the category of crossed split-plot designs are the two- and three-level factorial designs. Included within the category of non-crossed split-plots are the central composite design (CCD) and Box-Behnken designs. Two-level fractional factorial designs fall within the category of non-crossed designs when at least one sub-plot factor is confounded with an interaction involving a whole plot factor, and within the category of crossed split-plot designs otherwise.

An example of a crossed second order split-plot design with one whole plot factor and three whole plots is the 3^3 factorial design displayed in Figure 3.5a. The whole plot factor is displayed on the horizontal axis and the whole plots are indicated by means of a dotted ellipse. It is clear that the three whole plots all possess nine design points. Therefore, the design is said to be balanced. Moreover, the factor levels of the design points in the three whole plots only differ with respect to their whole plot factor level.

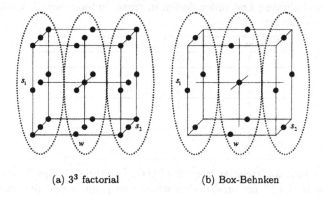

(a) 3^3 factorial   (b) Box-Behnken

Figure 3.5: Crossed and non-crossed split-plot design with one whole plot factor
and three whole plots.

Therefore, the design is crossed. An example of a non-crossed split-plot
design with one whole plot factor and three whole plots is given in Figure 3.5b. The design displayed is a Box-Behnken design. It is clear that the
split-plot design is not balanced and thus not crossed.

The equivalence of OLS and GLS in the case of crossed split-plot designs
has been shown by Letsinger et al. (1996). The proof is based on the fact
that

$$\mathbf{X'V^{-1}} = \mathbf{GX'},\tag{3.34}$$

where $\mathbf{G}$ is a lower triangular matrix depending on the design matrix $\mathbf{X}$.
Using (3.34), the GLS estimator becomes

$$(\mathbf{X'V^{-1}X})^{-1}\mathbf{X'V^{-1}y} = (\mathbf{GX'X})^{-1}\mathbf{GX'y},$$
$$= (\mathbf{X'X})^{-1}\mathbf{G^{-1}GX'y},$$
$$= (\mathbf{X'X})^{-1}\mathbf{X'y},$$

which is nothing but the OLS estimator. Crossed split-plot designs will be
revisited in Section 6.5.1, where their $\mathcal{D}$-optimality will be discussed.

3.5.4   First order non-crossed split-plot designs

Davison (1995) points out that GLS and OLS are also equivalent for fractional factorial designs provided that the sum of rows of the design matrix
within each whole plot is equal. If the levels are coded to -1 and +1, the
sum of the rows should be equal to a row vector of zeroes. Davison (1995)
erroneously conjectures that other fractions of two-level factorial designs
maintain the equivalency between OLS and GLS. As a matter of fact, con-

sider the following first order design in three factors, two of which are a whole plot factor:

Whole Plot	Factor 1	Factor 2	Factor 3
1	−1	−1	−1
2	−1	+1	−1
2	−1	+1	+1
3	+1	−1	−1
3	+1	−1	+1
4	+1	+1	−1
4	+1	+1	+1

This design is an irregular fraction of the 2^3 factorial design. More specifically, it is the 2^3 factorial design where one point is omitted. It can be shown that, for this design, OLS and GLS produce different results.

3.6   Small sample properties of the design criterion

The variance-covariance matrix (3.10) of the estimator $\hat{\beta}$ is asymptotic, whereas most industrial experiments have a limited sample size. In this section, we investigate whether the asymptotic variance-covariance matrix $X'V^{-1}X$ is a good approximation to the finite sample variance-covariance matrix. In case of an affirmative answer, the $\mathcal{D}$-criterion value $|X'V^{-1}X|$ can be considered a reliable measure of the quality of an experimental design. For our computations, we have distinguished two cases. Firstly, we assume that the variance components σ_ε^2 and σ_γ^2 have to be estimated. Secondly, we assume that an educated guess of σ_ε^2 and σ_γ^2 can be made based on prior information.

The 12 designs investigated are given in Table 3.8. The designs D1, D2, D4, D5, D6, D7, D8 and D9 are of the split-plot type, whereas the designs D3, D10, D11 and D12 are blocked designs. The designs D1, D2 and D3 are special because ordinary and generalized least squares estimation are equivalent for them. As a matter of fact, D1 and D2 are crossed split-plot designs, and D3 is an orthogonally blocked design with a homogeneous block size.

3.6.1   Estimating the variance components

We have conducted a simulation study to assess the impact of estimating the variance components on the information matrix of a finite sample experiment. For a number of standard response surface designs, we have performed 10,000 simulations in order to compute the finite sample information matrices and to compare them to their asymptotic counterparts. In each simulation run i, the unknown model parameter β was estimated by the generalized least squares estimate β^i using the REML estimates $\hat{\sigma}_\varepsilon^2$

Table 3.8: Designs used in the simulation study.

D1	3^3 factorial design with 1 whole plot variable.
D2	3^3 factorial design with 2 whole plot variables.
D3	Orthogonally blocked design given by Khuri (1994).
D4	3-variable CCD with $\alpha = 1.6818$, 6 center runs and 1 whole plot variable.
D5	3-variable CCD with $\alpha = 1.6818$, 6 center runs and 2 whole plot variables.
D6	4-variable CCD with $\alpha = 2$, 7 center runs and 1 whole plot variable.
D7	4-variable CCD with $\alpha = 2$, 7 center runs and 2 whole plot variables.
D8	3-variable Box-Behnken design with 3 center runs and 1 whole plot variable.
D9	3-variable Box-Behnken design with 3 center runs and 2 whole plot variables.
D10*	Face centered 4-variable CCD with duplicated axial points and 4 center runs, arranged in 9 blocks of size 4 by fractional partial confounding.
D11*	Face centered 4-variable CCD with duplicated axial points and 4 center runs, arranged in 9 blocks of size 4 by the algorithm of Trinca and Gilmour (1999).
D12*	Face centered 3-variable CCD with duplicated factorial points and 6 center runs, arranged in 7 blocks of size 4.

* The designs D10, D11 and D12 are displayed in Gilmour and Trinca (2000).

and $\hat{\sigma}_\gamma^2$ instead of σ_ε^2 and σ_γ^2. The variance-covariance matrix of all 10,000 β^i was computed in order to compare its trace and the inverse of its determinant to their asymptotic counterparts. In doing so, we have taken into account the stochastic nature of $\hat{\sigma}_\varepsilon^2$ and $\hat{\sigma}_\gamma^2$ in the computation of the finite sample information matrices.

In Table 3.9, the determinants of the asymptotic and the finite sample information matrix in a number of experimental situations are displayed. Determinant values for twelve different designs are given under a full quadratic model.

The simulation results for the designs D1, D2 and D3 serve as a benchmark. This is because, for these designs, the ordinary and generalized least squares estimates coincide. As a result, the asymptotic and the finite sample variance-covariance matrix are identical. In Table 3.9, this is reflected by the closeness of the determinants of the finite sample information matrices to the determinants of the corresponding asymptotic ones. For the other designs, the deviations are on average slightly higher. For the results displayed in Table 3.9, the mean absolute deviation amounts to 3.10% for the benchmark designs and to 7.03% for the other. In all experimental situations examined, the traces of the asymptotic and the finite sample information matrices are almost identical. This provides evidence that the asymptotic variance-covariance matrix may be considered as a close approximation to the finite sample covariance matrix in industrial experiments when the variance components are unknown. A similar conclusion was drawn in Letsinger et al. (1996). Therefore, the $\mathcal{D}$-optimality criterion, as defined in the previous section, can be considered a reliable

Table 3.9: Determinants of the asymptotic and finite sample information matrix for different designs and variance ratios.

	$\eta = 0.2$		$\eta = 1$		$\eta = 5$	
Design	ASYM	FINITE	ASYM	FINITE	ASYM	FINITE
D1	1.74E-04	1.66E-04	6.31E-04	5.96E-04	3.83E-01	3.61E-01
D2	2.24E-04	2.26E-04	1.54E-04	1.51E-04	2.22E-03	2.19E-03
D3	3.04E-03	2.94E-03	1.58E-01	1.53E-01	2.11E+03	2.12E+03
D4	8.78E-04	7.39E-04	1.05E-02	9.57E-03	1.28E+01	1.29E+01
D5	5.36E-04	5.08E-04	1.07E-03	1.05E-03	3.57E-02	3.52E-02
D6	2.09E-01	1.74E-01	2.75E+01	2.55E+01	8.26E+06	8.10E+06
D7	9.13E-02	7.62E-02	1.29E+00	1.18E+00	8.48E+03	8.32E+03
D8	2.09E-07	2.09E-07	1.35E-06	1.34E-06	1.00E-03	9.93E-04
D9	2.17E-07	2.23E-07	3.26E-07	3.32E-07	3.80E-06	3.82E-06
D10	3.87E-04	3.36E-04	6.02E-02	5.08E-02	5.07E+04	4.68E+04
D11	5.26E-04	4.34E-04	1.62E-01	1.40E-01	3.53E+05	3.17E+05
D12	9.49E-04	8.63E-04	3.81E-02	3.48E-02	4.47E+02	4.42E+02

statistic for comparing bi-randomization or two-stratum designs with a finite sample size when the variance components are estimated by REML.

3.6.2   Using an educated guess of the variance components

Often, information about the variance components is available from prior experimentation. Therefore, industrial experimenters may consider using an educated guess for the variance components rather than estimating them. Let us denote the resulting variance-covariance matrix of the observations by $\mathbf{V_e}$. The generalized least squares estimate of β then becomes

$$\hat{\beta} = (\mathbf{X'V_e^{-1}X})^{-1}\mathbf{X'V_e^{-1}y}.$$

This estimator is unbiased because

$$
\begin{aligned}
E(\hat{\beta}) &= E\{(\mathbf{X'V_e^{-1}X})^{-1}\mathbf{X'V_e^{-1}y}\}, \\
&= (\mathbf{X'V_e^{-1}X})^{-1}\mathbf{X'V_e^{-1}}\,E(\mathbf{y}), \\
&= (\mathbf{X'V_e^{-1}X})^{-1}\mathbf{X'V_e^{-1}}\{\mathbf{X}\beta + \mathbf{Z}\,E(\gamma) + E(\varepsilon)\}, \\
&= (\mathbf{X'V_e^{-1}X})^{-1}\mathbf{X'V_e^{-1}X}\beta, \\
&= \beta,
\end{aligned}
$$

and its finite sample variance-covariance matrix is given by

$$\text{var}(\hat{\beta}) = (\mathbf{X'V_e^{-1}X})^{-1}\mathbf{X'V_e^{-1}VV_e^{-1}X}(\mathbf{X'V_e^{-1}X})^{-1}. \qquad (3.35)$$

For D4 through D12, we have computed the finite sample information matrices, as given by the inverse of (3.35), assuming that an educated guess η_e of η was used. The more η_e deviates from η, the more the finite sample information matrix differs from the asymptotic one. This is illustrated in Figure 3.6, where the determinant of the finite sample information matrix

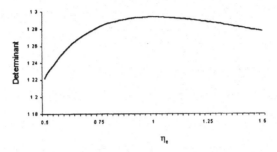

Figure 3.6: Determinants of the finite sample information matrix for design D7, obtained by using an educated guess η_e of the degree of correlation $\eta = 1$.

obtained by using an educated guess in design D7 is displayed. Table 3.10 contains the determinants for D4 through D12 in the cases where the degree of correlation was underestimated and overestimated by 50%. Comparing the determinants in Table 3.10 with the determinants of the corresponding asymptotic variance-covariance matrix in Table 3.9 shows that the differences are small. When the variance ratio is underestimated by 50%, the determinant of the resulting information matrix is on average 7.48% smaller than the determinant of the asymptotic information matrix. When the variance ratio is overestimated by 50%, it is only 4.26% smaller. The traces of the information matrices obtained by (3.35) are nearly identical to the asymptotic values. As a result, the asymptotic information matrix provides a close approximation to the finite sample information matrix when an educated guess of the variance ratio is used. The $\mathcal{D}$-criterion value is therefore a reliable statistic for comparing bi-randomization or two-stratum designs with a finite sample size when an educated guess is used for η.

Table 3.10: Determinants of finite sample information matrix for different designs and variance ratios when an educated guess η_e of η is used.

Design	$\eta = 0.2$		$\eta = 1$		$\eta = 5$	
	$\eta_e = 0.1$	$\eta_e = 0.3$	$\eta_e = 0.5$	$\eta_e = 1.5$	$\eta_e = 2.5$	$\eta_e = 7.5$
D4	8.55E-04	8.66E-04	1.00E-02	1.04E-02	1.26E+01	1.28E+01
D5	5.32E-04	5.33E-04	1.04E-03	1.06E-03	3.50E-02	3.55E-02
D6	1.99E-01	2.05E-01	2.68E+01	2.79E+01	8.16E+06	8.25E+06
D7	8.83E-02	8.98E-02	1.22E+00	1.28E+00	8.34E+02	8.47E+02
D8	2.09E-07	2.09E-07	1.35E-06	1.35E-06	1.00E-03	1.00E-03
D9	2.16E-07	2.16E-07	3.24E-07	3.25E-07	3.79E-06	3.80E-06
D10	3.68E-04	3.77E-04	5.16E-02	5.73E-02	4.31E+04	4.83E+04
D11	4.92E-04	5.08E-04	1.35E-01	1.56E-01	3.16E+05	3.47E+05
D12	9.15E-04	9.33E-04	3.49E-02	3.75E-02	4.27E+02	4.45E+02

4

Optimal Designs in the Presence of Random Block Effects

There are many experimental situations in which the experimental runs cannot be carried out under homogeneous conditions. For a more powerful analysis in these cases, the experiment is blocked so that each block is a homogeneous set of experimental units. In this chapter, it is assumed that the blocks are random and, hence, that the observations belonging to the same block are correlated.

4.1 Introduction

The raw material used in a production process is often obtained in batches in which the quality can vary considerably from one batch to another. To account for this variation among the batches, a random batch effect should be added to the regression model. In the semi-conductor industry, it is of interest to investigate the effect of several factors on the resistance in computer chips. Here, measurements are taken using silicon wafers randomly drawn from a large lot. Therefore, the wafer effect should be considered as a random effect in the corresponding model. In Example 4 of Section 3.1, an optometry experiment is described, in which a response is measured for each eye of the test subjects. As a result, each subject's pair of eyes is a block of two possibly correlated observations. Other examples of experiments where there might be random block effects include agricultural experiments where multiple fields are used or chemistry experiments where runs are executed on different days or in different laboratories. The experimental design question in these examples is how to allocate the levels of the factors under investigation to the blocks. Although there exists an extensive literature on the blocking of designs for qualitative factors, optimal block designs for quantitative factors have received much less attention. Atkinson and Donev (1989) and Cook and Nachtsheim (1989) propose an exchange algorithm for the computation of $\mathcal{D}$-optimal regression designs in the presence of fixed block effects. The derivation of continuous optimal regression designs in the presence of random block effects has been studied by Atkins (1994), Cheng (1995) and by Atkins and Cheng (1999). However, their approximate theory for the design of blocked experiments is of limited practical use in industrial environments where the number of blocks is typically small. Cheng (1995) as well as Atkins and Cheng (1999) also describe a special case in which the exact optimal design is easy to construct. Chasalow (1992) uses complete enumeration to find exact designs for quadratic regression when there are random block effects. For more complicated models, complete enumeration becomes impossible within a reasonable computation time. In this chapter, we describe the exchange algorithm of Goos and Vandebroek (2001a) for this design problem. The analysis of response surface models with random block effects is discussed by Khuri (1992), who also derives general conditions for orthogonal blocking.

4.2   The random block effects model

Assume that an experiment consists of n experimental runs arranged in b blocks of sizes $k_1, k_2, \ldots, k_b$ with $n = \sum_{i=1}^{b} k_i$. When the blocks are random, the response y_{ij} of the jth observation in the ith block can be written as

$$y_{ij} = \mathbf{f}'(\mathbf{x}_{ij})\boldsymbol{\beta} + \gamma_i + \varepsilon_{ij}, \tag{4.1}$$

where $\mathbf{f}$ is the polynomial expansion of the explanatory variables $\mathbf{x}$, $\boldsymbol{\beta} = [\,\beta_1\ \beta_2\ \ldots\ \beta_p\,]'$ is the $p \times 1$ vector containing the effects of the experimental variables, γ_i represents the ith block effect and ε_{ij} is the random error. In matrix notation, the model becomes

$$\mathbf{y} = \mathbf{X}\boldsymbol{\beta} + \mathbf{Z}\boldsymbol{\gamma} + \boldsymbol{\varepsilon}, \tag{4.2}$$

where $\mathbf{y}$ is a vector of n observations on the response of interest, the vector $\boldsymbol{\beta}$ contains the p unknown fixed parameters, the vector $\boldsymbol{\gamma} = [\,\gamma_1\ \gamma_2\ \ldots\ \gamma_b\,]'$ contains the b random block effects and $\boldsymbol{\varepsilon}$ is a random error vector. The matrices $\mathbf{X}$ and $\mathbf{Z}$ are known and have dimension $n \times p$ and $n \times b$ respectively. $\mathbf{X}$ contains the polynomial expansions of the m factor levels at the n experimental runs. $\mathbf{Z}$ is of the form

$$\mathbf{Z} = \mathrm{diag}[\mathbf{1}_{k_1}, \mathbf{1}_{k_2}, \ldots, \mathbf{1}_{k_b}], \tag{4.3}$$

where $\mathbf{1}_k$ is a $k \times 1$ vector of ones. It is assumed that

$$E(\boldsymbol{\varepsilon}) = \mathbf{0}_n \text{ and } \mathrm{Cov}(\boldsymbol{\varepsilon}) = \sigma_\varepsilon^2 \mathbf{I}_n, \tag{4.4}$$

$$E(\boldsymbol{\gamma}) = \mathbf{0}_b \text{ and } \mathrm{Cov}(\boldsymbol{\gamma}) = \sigma_\gamma^2 \mathbf{I}_b, \tag{4.5}$$

$$\text{and } \mathrm{Cov}(\boldsymbol{\gamma}, \boldsymbol{\varepsilon}) = \mathbf{0}_{b \times n} \tag{4.6}$$

We will refer to this model as the random block effects model. In Chapter 3, it is shown that the variance-covariance matrix of the observations $\mathbf{y}$ is compound symmetric and that the information matrix on the unknown model parameters $\boldsymbol{\beta}$ is equal to

$$\mathbf{M} = \mathbf{X}'\mathbf{V}^{-1}\mathbf{X},$$

$$= \frac{1}{\sigma_\varepsilon^2}\left\{\mathbf{X}'\mathbf{X} - \sum_{i=1}^{b}\frac{\eta}{1+k_i\eta}(\mathbf{X}_i'\mathbf{1}_{k_i})(\mathbf{X}_i'\mathbf{1}_{k_i})'\right\}, \tag{4.7}$$

where η is the degree of correlation. When $\eta \to 0$, or equivalently $\sigma_\gamma^2 \to 0$, the observations within the same block are no longer correlated and the random block effects model degenerates to the uncorrelated model (1.3). The information matrix then becomes

$$\mathbf{M} = \frac{1}{\sigma_\varepsilon^2}\mathbf{X}'\mathbf{X}. \tag{4.8}$$

This result suggests that an optimal design for the random block effects model is likely to have the same design points as an optimal design for the uncorrelated model when η is small. From the information matrix (4.7), it can easily be seen that it also reduces to (4.8) when $\mathbf{X}_i'\mathbf{1}_{k_i} = \mathbf{0}_p$ ($i = 1, 2, \ldots, b$). In that case, the information matrix does not depend on the degree of correlation η and the optimal design points are also given by the optimal design for the uncorrelated model (1.3). We will pay attention to this and other cases in which the optimal design points do not depend on η in Section 4.3. In Section 4.4, we investigate the case where the degree of correlation η becomes infinitely large, that is $\eta \to \infty$. We show that the problem of designing an experiment with random block effects is then equivalent to that of designing an experiment with fixed block effects. Next, we concentrate on the general case where the optimal design depends on the degree of correlation η. We develop an algorithm to compute $\mathcal{D}$-optimal designs and describe the computational results. In the first instance, we restrict ourselves to three different levels for each experimental factor. In the second instance, we investigate to what extent the three-level designs can be improved by allowing more factor levels. Finally, we show that orthogonal blocking is an optimal design strategy for a given design matrix $\mathbf{X}$.

4.3   Optimal designs that do not depend on η

In four specific cases, the $\mathcal{D}$-optimal design for model (4.2) does not depend on the degree of correlation η. In each case, the optimal block design is based on an optimal design for the uncorrelated model

$$\mathbf{y} = \mathbf{X}\beta + \varepsilon, \tag{4.9}$$

where $\operatorname{Cov}(\varepsilon) = \sigma_\varepsilon^2 \mathbf{I}_n$ and $\mathbf{y}$, $\mathbf{X}$ and β are defined as in the correlated model (4.2).

4.3.1   Orthogonally blocked first order designs

An orthogonal block design that is supported on the points of a $\mathcal{D}$-optimal design for the uncorrelated model (4.9) with

$$\mathbf{X}_i' \mathbf{1}_{k_i} = \mathbf{0}_p, \qquad (i = 1, 2, \dots, b), \tag{4.10}$$

is a $\mathcal{D}$-optimal design for model (4.2). The design so obtained is optimal for any positive η, so that no prior knowledge of the variance components is needed. Note that the block size may be heterogeneous. Condition (4.10) is a special case of the general conditions for orthogonal blocking of response surface designs given in Section 2.3.4.

Assume that the observations are arranged so that (4.10) holds. The information matrix (4.7) then simplifies to $\mathbf{M}_{\mathrm{orth.}} = \mathbf{X}'\mathbf{X}/\sigma_\varepsilon^2$, which is the information matrix on the unknown parameters in model (4.9). When the observations are arranged so that (4.10) does not hold, at least one $\mathbf{X}_i' \mathbf{1}_{k_i} \neq \mathbf{0}$ and the information matrix is of the form (4.7). The difference

$$\mathbf{M}_{\mathrm{orth.}} - \mathbf{M} = \frac{1}{\sigma_\varepsilon^2}\left\{ \sum_{i=1}^b \frac{\eta}{1 + k_i \eta}(\mathbf{X}_i' \mathbf{1}_{k_i})(\mathbf{X}_i' \mathbf{1}_{k_i})' \right\}$$

is then a nonnegative definite matrix. Therefore, blocking an experiment so that (4.10) holds is an optimal design strategy for a given design matrix $\mathbf{X}$ independent of the design criterion used.

If, in addition, this design matrix $\mathbf{X}$ maximizes $|\mathbf{X}'\mathbf{X}/\sigma_\varepsilon^2|$, then the resulting block design is $\mathcal{D}$-optimal for the random block effects model. Now, a $\mathcal{D}$-optimal design for the uncorrelated model (4.9) maximizes $|\mathbf{X}'\mathbf{X}/\sigma_\varepsilon^2|$. Therefore, arranging the n observations of a $\mathcal{D}$-optimal design for the uncorrelated model (4.9) in blocks so that (4.10) holds is an optimal design strategy for the random block effects model.

When the model of interest contains an intercept, the design matrix $\mathbf{X}$ contains a column of ones so that (4.10) cannot be satisfied. However, we can show that a blocked design is $\mathcal{D}$-optimal when

$$\mathbf{X}_i' \mathbf{1}_{k_i} = \begin{pmatrix} k_i \\ \mathbf{0}_{p-1} \end{pmatrix}, \qquad (i = 1, 2, \ldots, b), \tag{4.11}$$

provided the first column of $\mathbf{X}$ corresponds to the intercept. Equivalently, this condition can be written as

$$\tilde{\mathbf{X}}_i' \mathbf{1}_{k_i} = \mathbf{0}_{p-1}, \qquad (i = 1, 2, \ldots, b), \tag{4.12}$$

where $\mathbf{X} = [\mathbf{1}_n \quad \tilde{\mathbf{X}}]$. Using this notation, the information matrix on $\boldsymbol{\beta}$ is given by

$$\mathbf{X}' \mathbf{V}^{-1} \mathbf{X} = \begin{bmatrix} \mathbf{1}_n' \mathbf{V}^{-1} \mathbf{1}_n & \mathbf{1}_n' \mathbf{V}^{-1} \tilde{\mathbf{X}} \\ \tilde{\mathbf{X}}' \mathbf{V}^{-1} \mathbf{1}_n & \tilde{\mathbf{X}}' \mathbf{V}^{-1} \tilde{\mathbf{X}} \end{bmatrix}.$$

The $\mathcal{D}$-optimal design maximizes the determinant of the information matrix. From Theorem 13.3.8 in Harville (1997), we have that

$$|\mathbf{X}' \mathbf{V}^{-1} \mathbf{X}| = (\mathbf{1}_n' \mathbf{V}^{-1} \mathbf{1}_n) |\tilde{\mathbf{X}}' \mathbf{V}^{-1} \tilde{\mathbf{X}} - \tilde{\mathbf{X}}' \mathbf{V}^{-1} \mathbf{1}_n (\mathbf{1}_n' \mathbf{V}^{-1} \mathbf{1}_n)^{-1} \mathbf{1}_n' \mathbf{V}^{-1} \tilde{\mathbf{X}}|.$$

Since $c_1 = \mathbf{1}_n' \mathbf{V}^{-1} \mathbf{1}_n = \sigma_\varepsilon^{-2} \sum_{i=1}^{b} k_i / (1 + k_i \eta)$ is constant over all possible designs, the $\mathcal{D}$-optimality criterion only depends on

$$\tilde{\mathbf{X}}' \mathbf{V}^{-1} \tilde{\mathbf{X}} - \tilde{\mathbf{X}}' \mathbf{V}^{-1} \mathbf{1}_n (\mathbf{1}_n' \mathbf{V}^{-1} \mathbf{1}_n)^{-1} \mathbf{1}_n' \mathbf{V}^{-1} \tilde{\mathbf{X}}. \tag{4.13}$$

As a result, the $\mathcal{D}$-optimal design for estimating $\boldsymbol{\beta}$ is equivalent to the $\mathcal{D}_s$- or $\mathcal{D}_{\tilde{\beta}}$-optimal design for estimating $\tilde{\boldsymbol{\beta}}$. Letting $c_{2i} = 1/(1 + k_i \eta)$ for $i = 1, 2, \ldots, b$ and substituting

$$\tilde{\mathbf{X}}' \mathbf{V}^{-1} \tilde{\mathbf{X}} = \sigma_\varepsilon^{-2} \left\{ \tilde{\mathbf{X}}' \tilde{\mathbf{X}} - \sum_{i=1}^{b} \frac{\eta}{1 + k_i \eta} (\tilde{\mathbf{X}}_i' \mathbf{1}_{k_i})(\tilde{\mathbf{X}}_i' \mathbf{1}_{k_i})' \right\},$$

and

$$\tilde{\mathbf{X}}'\mathbf{V}^{-1}\mathbf{1}_n = \sum_{i=1}^{b} \tilde{\mathbf{X}}_i'\mathbf{V}_i^{-1}\mathbf{1}_{k_i},$$

$$= \sigma_\varepsilon^{-2}\sum_{i=1}^{b}\{\tilde{\mathbf{X}}_i'(\mathbf{I}_{k_i} - \frac{\eta}{1+k_i\eta}\tilde{\mathbf{X}}_i'\mathbf{1}_{k_i}\mathbf{1}_{k_i}')\mathbf{1}_{k_i}\},$$

$$= \sigma_\varepsilon^{-2}\sum_{i=1}^{b}(\tilde{\mathbf{X}}_i'\mathbf{1}_{k_i} - \frac{\eta}{1+k_i\eta}\tilde{\mathbf{X}}_i'\mathbf{1}_{k_i}\mathbf{1}_{k_i}'\mathbf{1}_{k_i}),$$

$$= \sigma_\varepsilon^{-2}\sum_{i=1}^{b}(\tilde{\mathbf{X}}_i'\mathbf{1}_{k_i} - \frac{k_i\eta}{1+k_i\eta}\tilde{\mathbf{X}}_i'\mathbf{1}_{k_i}),$$

$$= \sigma_\varepsilon^{-2}\sum_{i=1}^{b}\frac{\tilde{\mathbf{X}}_i'\mathbf{1}_{k_i}}{1+k_i\eta},$$

$$= \sigma_\varepsilon^{-2}\sum_{i=1}^{b}c_{2i}\tilde{\mathbf{X}}_i'\mathbf{1}_{k_i},$$

in (4.13) yields

$$\tilde{\mathbf{X}}'\tilde{\mathbf{X}} - \sum_{i=1}^{b}\frac{\eta}{1+k_i\eta}(\tilde{\mathbf{X}}_i'\mathbf{1}_{k_i})(\tilde{\mathbf{X}}_i'\mathbf{1}_{k_i})' - c_1^{-1}\sigma_\varepsilon^{-2}(\sum_{i=1}^{b}c_{2i}\tilde{\mathbf{X}}_i'\mathbf{1}_{k_i})(\sum_{i=1}^{b}c_{2i}\tilde{\mathbf{X}}_i'\mathbf{1}_{k_i})',$$

$$(4.14)$$

apart from the constant σ_ε^{-2}. When condition (4.12) holds, this matrix reduces to

$$\tilde{\mathbf{X}}'\tilde{\mathbf{X}}. \qquad (4.15)$$

As a result, the difference between the matrices (4.15) and (4.14) is

$$\sum_{i=1}^{b}\frac{\eta}{1+k_i\eta}(\tilde{\mathbf{X}}_i'\mathbf{1}_{k_i})(\tilde{\mathbf{X}}_i'\mathbf{1}_{k_i})' + c_1^{-1}\sigma_\varepsilon^{-2}(\sum_{i=1}^{b}c_{2i}\tilde{\mathbf{X}}_i'\mathbf{1}_{k_i})(\sum_{i=1}^{b}c_{2i}\tilde{\mathbf{X}}_i'\mathbf{1}_{k_i})'.$$

This matrix difference is nonnegative definite because it is a sum of outer products of vectors. As a result, the determinant of (4.15) is larger than that of (4.14) as soon as one $\tilde{\mathbf{X}}_i'\mathbf{1}_{k_i} \neq \mathbf{0}_{p-1}$. Therefore, assigning observations to the blocks so that (4.10) holds is the optimal strategy for a given $\tilde{\mathbf{X}}$. In addition, a $\mathcal{D}$-optimal design for the uncorrelated model (4.9) maximizes

$$|\mathbf{X}'\mathbf{X}| = (\mathbf{1}_n'\mathbf{1}_n)|\tilde{\mathbf{X}}'\tilde{\mathbf{X}} - \tilde{\mathbf{X}}'\mathbf{1}_n(\mathbf{1}_n'\mathbf{1}_n)^{-1}\mathbf{1}_n'\tilde{\mathbf{X}}|,$$
$$= n|\tilde{\mathbf{X}}'\tilde{\mathbf{X}} - n^{-1}(\tilde{\mathbf{X}}'\mathbf{1}_n)(\tilde{\mathbf{X}}'\mathbf{1}_n)'|,$$

which simplifies to $n|\tilde{\mathbf{X}}'\tilde{\mathbf{X}}|$ when (4.12) holds. Therefore, an orthogonal design that is supported on the $n = \sum_{i=1}^{b} k_i$ points of a $\mathcal{D}$-optimal design for the uncorrelated model (4.9) so that (4.12), or equivalently (4.11), holds, is a $\mathcal{D}$-optimal design for model (4.2).

Block 1			Block 2		
A	B	C	A	B	C
−1	−1	−1	+1	−1	−1
+1	+1	−1	−1	+1	−1
+1	−1	+1	−1	−1	+1
−1	+1	+1	+1	+1	+1

Table 4.1: $\mathcal{D}$-optimal block design for the whipped topping experiment.

The conditions can be generalized because the $\mathcal{D}$-optimality criterion is invariant to a linear transformation of the factor levels. Therefore, any design for which (4.10) and (4.11) can be accomplished by applying a linear transformation on the factor levels is $\mathcal{D}$-optimal for the problem at hand.

As a first illustration, consider an experiment from Cook and Nachtsheim (1989) to optimize the stability of a whipped topping. The amount of two emulsifiers (A and B) and the amount of fat (C) are expected to have an impact on the melting that occurs after the aerosol topping is dispensed. Suppose that the experimenters are interested in the linear effects and the two factor interactions only and that two laboratory assistants are available for eight observations. The familiar 2^3 factorial is a $\mathcal{D}$-optimal design with eight observations for estimating the effects of interest in the uncorrelated model. A $\mathcal{D}$-optimal design in the presence of random block effects is easily obtained by using ABC as the block generator (for more details on blocking 2^m factorial designs, we refer the reader to Chapter 9). The resulting design is displayed in Table 4.1. It is easy to verify that condition (4.10) holds.

As another illustration, consider an experiment with 32 observations to estimate the intercept, the five linear effects and eight two-factor interactions. Suppose the model under investigation is given by

$$y = \beta_0 + \sum_{i=1}^{5} \beta_i x_i + \sum_{i=1}^{5}\sum_{j=i}^{5} \beta_{ij} x_i x_j, \qquad (4.16)$$
$$(i,j) \neq (1,3)$$
$$(i,j) \neq (2,4)$$

and that the observations should be assigned to four blocks of four observations and two blocks of eight observations. An optimal design is displayed in Table 4.2. It will not come as a surprise that the optimal design points are given by the 2^5 factorial design. An optimal assignment to the blocks is obtained in two steps. Firstly, the design points are arranged in 8 blocks of 4 observations by using the block generators ABE, BCE and CDE. Secondly, two pairs of blocks are selected and merged into two blocks of size eight. The results of Section 4.3.1 can be used to verify that the design obtained in this way is indeed optimal.

Block	Factor levels					Block	Factor levels				
	A	B	C	D	E		A	B	C	D	E
1	-1	-1	-1	-1	-1	3	-1	1	-1	-1	-1
	1	1	1	1	-1		1	-1	1	1	-1
	1	-1	1	-1	1		1	1	1	-1	1
	-1	1	-1	1	1		-1	-1	-1	1	1
	1	-1	1	-1	-1	4	1	1	-1	-1	-1
	-1	1	-1	1	-1		-1	1	1	-1	1
	-1	-1	-1	-1	1		-1	-1	1	1	-1
	1	1	1	1	1		1	-1	-1	1	1
2	1	-1	-1	-1	-1	5	1	-1	1	1	1
	-1	1	1	1	-1		-1	1	-1	-1	1
	-1	-1	1	-1	1		1	1	1	-1	-1
	1	1	-1	1	1		-1	-1	-1	1	-1
	-1	-1	1	-1	-1	6	-1	1	1	-1	-1
	1	1	-1	1	-1		1	-1	-1	1	-1
	1	-1	-1	-1	1		-1	-1	1	1	1
	-1	1	1	1	1		1	1	-1	-1	1

Table 4.2: $\mathcal{D}$-optimal block design with two blocks of eight observations and four blocks of four observations for the estimation of model (4.16).

It is clear that conditions (4.10) and (4.11) cannot be satisfied when quadratic terms are included in the model. The same goes for pure linear models and linear models with interactions when the block size is an odd number. This is because a $\mathcal{D}$-optimal design for the uncorrelated pure linear model or the linear model with interactions has factor levels -1 and +1 only.

4.3.2   Saturated designs

It is easy to see that the $\mathcal{D}$-optimal saturated design for the random block effects model does not depend on the degree of correlation η. The optimal design points are those from a $\mathcal{D}$-optimal saturated design for the uncorrelated model (4.9).

A saturated design is a design where the number of observations n is equal to the number of unknown parameters p. As a result, its design matrix $\mathbf{X}$ is a $p \times p$ matrix and its $\mathcal{D}$-criterion value becomes

$$|\mathbf{X}'\mathbf{V}^{-1}\mathbf{X}| = |\mathbf{X}'|\,|\mathbf{V}^{-1}|\,|\mathbf{X}|,$$
$$= |\mathbf{V}^{-1}|\,|\mathbf{X}'|\,|\mathbf{X}|, \qquad (4.17)$$
$$= |\mathbf{V}|^{-1}|\mathbf{X}'\mathbf{X}|.$$

Since we assume that the block size is dictated by the experimental situation, $|\mathbf{V}|$ is a constant for all design options. Hence, the $\mathcal{D}$-optimal saturated design for the random block effects model can be found by maximizing $|\mathbf{X}'\mathbf{X}|$. This is nothing else than determining a $\mathcal{D}$-optimal design for the uncorrelated model (see Section 1.4.2). From (4.17), it can be seen that the assignment of observations to the blocks does not influence the $\mathcal{D}$-optimality criterion. Suppose for example that we are interested in the estimation of the full second order model in two variables:

$$y = \beta_0 + \beta_1 x_1 + \beta_2 x_2 + \beta_{12} x_1 x_2 + \beta_{11} x_1^2 + \beta_{22} x_2^2. \qquad (4.18)$$

This model has six unknown parameters. The six factor level combinations of the $\mathcal{D}$-optimal saturated design are displayed in Figure 1.2a. If two blocks of three observations are available, then a $\mathcal{D}$-optimal design for the random block effects model is obtained by randomly assigning the six design points to the blocks:

Block	x_1	x_2
	-1	1
1	3α	1
	1	3α
	-1	-1
2	$-\alpha$	$-\alpha$
	1	-1

where $\alpha = (4 - \sqrt{13})/3$.

4.3.3   Product designs

In a few specific instances, a $\mathcal{D}$- and $\mathcal{A}$-optimal design for the correlated model (4.2) can be constructed from the $\mathcal{D}$- and $\mathcal{A}$-optimal continuous design for the uncorrelated model (4.9) when the block size is larger than the number of model parameters. A similar result was found for the fixed block effects model by Kurotschka (1981). More details can be found in Section 2.3.8.

Let $\{\mathbf{x}_1, \mathbf{x}_2, \ldots, \mathbf{x}_h\}$ with weights $\{w_1, w_2, \ldots, w_h\}$ be a $\mathcal{D}$- or $\mathcal{A}$-optimal continuous design for the uncorrelated model (4.9). Atkins and Cheng (1999) prove that the $\mathcal{D}$- and $\mathcal{A}$-optimal design for the random block effects model (4.2) with b blocks of size k consists of b identical blocks where each design point $\mathbf{x}_i$ is replicated $k w_i$ times if $k w_i$ is an integer for each i. The optimal design is then independent of η. This theorem is only valid when the model contains an intercept. It implicitly requires that the block size k is greater than or equal to the number of parameters p. As a matter of fact, each block is supported on at least h distinct design points, so that $k \geq h$. In addition, the number of distinct design points h must be greater than or equal to the number of parameters p in order to have a nonsingular design. It is interesting to note that this type of design is orthogonally blocked. In fact, all blocks are equal so that $n = bk$ and $\mathbf{X}' \mathbf{1}_n = b \mathbf{X}_i' \mathbf{1}_{k_i}$. Substituting this in (2.39) shows that the orthogonality condition is satisfied.

The theorem has a serious impact on the design of this type of experiments. Firstly, no prior knowledge on η is required since the optimal design points only depend on model (4.9). Secondly, rather than using a computationally intensive blocking algorithm to generate a design with $n = bk$ observations, a k-point $\mathcal{D}$- or $\mathcal{A}$-optimal design for the uncorrelated model can be used in each block. Although occasions in which all kw_i are integer are rare, the result of Atkins and Cheng (1999) is expected to be useful when the block size k is large with respect to the number of parameters. In this case, the values kw_i can be rounded to the nearest integer without serious loss of design efficiency. Unfortunately, design problems for which $k \geq p$ seldom occur in practice.

Well-known situations where all kw_i can be integer occur in mixture experiments and in the case of quadratic regression on a single explanatory variable. In mixture experiments, the optimal continuous designs for first and second order models have equal weight on the s points of a simplex lattice design. $\mathcal{D}$-optimal block designs when k is a multiple of s consist of identical blocks in which the lattice design is replicated k/s times. The attentive reader will point out that mixture models typically do not contain an intercept and that Atkins and Cheng's theorem does not apply in that case. However, a linear transformation of the design matrix of a mixture experiment exists so that it does contain a column of ones. This is because the sum of the mixture components always equals one. For quadratic regression on one variable, the $\mathcal{D}$-optimal continuous design has weight $1/3$ on the points -1, 0 and 1. Therefore, a $\mathcal{D}$-optimal design for quadratic regression on $[-1, 1]$ in the presence of random block effects can be readily obtained when the block size is a multiple of three. Consider an experiment carried out to investigate the impact of the initial potassium/carbon (K/C) ratio on the desorption of carbon monoxide (CO) in the context of the gasification of coal. The experiment is described in Atkinson and Donev (1992). Let x and y denote the K/C ratio and the amount of CO desorbed respectively. Further, suppose that four blocks of observations are available to the researcher and that the model of interest is given by

$$y = \beta_0 + \beta_1 x + \beta_{11} x^2. \tag{4.19}$$

If the block size of the experiment is equal to three, a $\mathcal{D}$-optimal design with four blocks of three observations consists of four identical blocks in which -1, 0 and 1 each appear once. Similarly, a $\mathcal{D}$-optimal design with four blocks of six observations each consists of four identical blocks in which -1, 0 and 1 each appear twice. In a similar way, $\mathcal{A}$-optimal block designs for the coal gasification experiment can be constructed when the block size is a multiple of four. For example, an $\mathcal{A}$-optimal design with four blocks of eight observations has four identical blocks in which both -1 and 1 appear twice and 0 is replicated four times. This is because the $\mathcal{A}$-optimal continuous design for quadratic regression on $[-1, 1]$ has weight $1/4$ on -1 and 1 and weight $1/2$ on 0.

4.3.4 Minimum support designs

In this section, we restrict our attention to minimum support designs. A minimum support design for a model with p parameters is supported on exactly p distinct points $\mathbf{x}_1, \mathbf{x}_2, \ldots, \mathbf{x}_p$. This class of designs is useful because experimenters are often in favor of using small numbers of different factor levels and design points. Using a smaller number of distinct points than p in the experiment would result in a singular information matrix. Cheng (1995) provides a method to construct $\mathcal{D}$-optimal minimum support designs for the random block effects model (4.2) by combining a p-point $\mathcal{D}$-optimal design for the uncorrelated model (4.9) and a design with b blocks of size k for p treatments that is $\mathcal{D}$-optimal for estimating treatment contrasts. The most famous examples of $\mathcal{D}$-optimal designs for comparing treatments are balanced incomplete block designs (see Section 1.6.2).

Suppose $\{\mathbf{x}_1, \mathbf{x}_2, \ldots, \mathbf{x}_p\}$ is a $\mathcal{D}$-optimal design with p observations for the uncorrelated model (4.9) and that a balanced incomplete block design exists with p treatments and b blocks of size k, then using the p design points as treatments in the balanced incomplete block design yields a design that is $\mathcal{D}$-optimal in the class of block designs with p points and b blocks of size k. As in the Sections 4.3.1 through 4.3.3, no prior knowledge on η is required and only a small design for the uncorrelated model has to be computed. The main drawback of this approach is that the block size should be homogeneous.

As an illustration, consider a modified version of the constrained mixture experiment for estimating the impact of three factors on the electric resistivity of a modified acrylonitrile powder described in Atkinson and Donev (1992). The factors under investigation are

x_1 copper sulphate (CuSO$_4$),
x_2 sodium thiosulphate (Na$_2$S$_2$O$_3$),
x_3 glyoxal (CHO)$_2$.

The following constraints were imposed on the factor levels:

$$
\begin{aligned}
0.2 &\leq x_1 \leq 0.8, \\
0.2 &\leq x_2 \leq 0.8, \\
0.0 &\leq x_3 \leq 0.6.
\end{aligned}
$$

Assume that the model is given by the second-order Scheffé polynomial

$$y = \sum_{i=1}^{3} \beta_i x_i + \sum_{i=1}^{2} \sum_{j=i+1}^{3} \beta_{ij} x_i x_j. \tag{4.20}$$

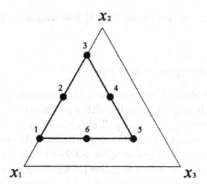

Figure 4.1: $\mathcal{D}$-optimal design for the constrained mixture experiment.

Table 4.3: Balanced incomplete block design with six treatments and ten blocks of size three.

Block	Treatments			Block	Treatments		
1	1	2	5	6	2	3	4
2	1	2	6	7	2	3	5
3	1	3	4	8	2	4	6
4	1	3	6	9	3	5	6
5	1	4	5	10	4	5	6

A $\mathcal{D}$-optimal design with six observations for this model is given by the second order simplex lattice design on the constrained design region. This is illustrated in Figure 4.1. Now, suppose that 10 experimenters are available and that 30 observations are considered desirable. Combining the $\mathcal{D}$-optimal design in Figure 4.1 and a balanced incomplete block design for six treatments and ten blocks of size three then yields an optimal design in the class of minimum support designs. The balanced incomplete block design is shown in Table 4.3. The experiment is carried out by using each of the six design points as a treatment in the balanced incomplete block design.

In general, minimum support designs will be less efficient than overall optimal designs. For example, a minimum support design with three blocks of size three for the estimation of a full quadratic model in two variables is only 76.3% as efficient as the overall $\mathcal{D}$-optimal design when $\eta = 0.1$. For $\eta = 10$, the minimum support design is only 69.4% as efficient. As a result, a serious loss of information can be incurred by restricting to minimum support designs.

4.4   Optimal designs when $\eta \to +\infty$

In this section, we show that the $\mathcal{D}$-optimal design for the random block effects model when η is infinitely large is equivalent to the $\mathcal{D}$-optimal design for the fixed block effects model. In Section 2.3.6, we found that the $\mathcal{D}$-optimal design for the latter model maximizes the determinant of

$$\tilde{\mathbf{X}}'\tilde{\mathbf{X}} - \sum_{i=1}^{b} \frac{1}{k_i}(\tilde{\mathbf{X}}_i'\mathbf{1}_{k_i})(\tilde{\mathbf{X}}_i'\mathbf{1}_{k_i})'. \tag{4.21}$$

In Section 4.3.1, we have shown that the $\mathcal{D}$-optimal design for estimating β is equivalent to the $\mathcal{D}_s$- or $\mathcal{D}_{\tilde{\beta}}$-optimal design for estimating $\tilde{\beta}$. As a consequence, the $\mathcal{D}$-optimal design for the random block effects model maximizes the determinant of

$$\tilde{\mathbf{X}}'\tilde{\mathbf{X}} - \sum_{i=1}^{b} \frac{\eta}{1+k_i\eta}(\tilde{\mathbf{X}}_i'\mathbf{1}_{k_i})(\tilde{\mathbf{X}}_i'\mathbf{1}_{k_i})' - c_1^{-1}\sigma_\epsilon^{-2}(\sum_{i=1}^{b} c_{2i}\tilde{\mathbf{X}}_i'\mathbf{1}_{k_i})(\sum_{i=1}^{b} c_{2i}\tilde{\mathbf{X}}_i'\mathbf{1}_{k_i})', \tag{4.22}$$

where $c_1 = \sigma_\epsilon^{-2}\sum_{i=1}^{b} k_i/(1+k_i\eta)$ and $c_{2i} = 1/(1+k_i\eta)$ $(i = 1,2,\ldots,b)$. When $\eta \to +\infty$, this expression reduces to

$$\tilde{\mathbf{X}}'\tilde{\mathbf{X}} - \sum_{i=1}^{b} \frac{1}{k_i}(\tilde{\mathbf{X}}_i'\mathbf{1}_{k_i})(\tilde{\mathbf{X}}_i'\mathbf{1}_{k_i})', \tag{4.23}$$

which is identical to (4.21). As a consequence, the $\mathcal{D}$-optimal designs for the random block effects model is equivalent to the $\mathcal{D}$-optimal design for the fixed block effects model when η is infinitely large.

This result implies that the design problem at hand is related to that considered in Atkinson and Donev (1989) and Cook and Nachtsheim (1989), namely the design of experiments for the fixed block effects model. It is true that blocked experiments generated for fixed block effects models can be used when the blocks are random as well. We have shown that this makes sense if η is large. However, it is expected that these designs will not be optimal for practical values of η. The algorithms based on the assumption of fixed block effects also fail to produce designs when $p + b > n$ because b block effects need to be estimated when the blocks are fixed rather than random. For these reasons, we have developed an algorithm to compute $\mathcal{D}$-optimal designs in the presence of random block effects. Designs can be produced as soon as $n \geq p$.

4.5   The general case

Many experimental situations exist where the optimal design depends on the degree of correlation η. As a matter of fact, the results of Section 4.3

are applicable only in a limited number of cases. For example, the result on orthogonally blocked first order designs can only be used when the block size is an even number and the theorem on minimum support designs in Section 4.3.4 cannot be used when the number of support points is allowed to be larger than the number of unknown parameters. In this section, we develop a design construction algorithm for the general case where the $\mathcal{D}$-optimal design depends on the value of η.

4.5.1   Complete enumeration

Chasalow (1992) presents a complete enumeration approach to find discrete optimal designs for the random block effects model and applies it to the case of quadratic regression on $[-1, 1]$. This approach involves enumerating all possible blocks of the appropriate size as well as all possible designs consisting of these blocks. Therefore, it is computationally intensive when more than one factor is under investigation. Suppose, for example, that an experiment with 6 blocks of 4 observations is conducted to estimate a full quadratic model in 2 variables with 3 factor levels. Since we have $3^2 = 9$ factor level combinations, the number of different blocks of 4 observations is given by

$$\binom{4 + 9 - 1}{4} = 495.$$

As a result, the total number of designs considered is given by

$$\binom{6 + 495 - 1}{6} \sim 10^{13}.$$

It is clear that increasing the number of experimental variables or the number of factor levels would further complicate the search for an optimal design, as well as the presence of heterogeneous block sizes.

4.5.2   Generic point exchange algorithm

Unlike Chasalow (1992), we have chosen to use a point exchange algorithm to compute $\mathcal{D}$-optimal designs under random block effects. This is because enumerating all possible blocks and designs is a hopeless task when two or more factors are under investigation and when more than three factor levels are considered. Point exchange algorithms have been used for a variety of design problems, one of them being the blocking of response surface designs when the block effects are fixed. This topic is treated in Atkinson and Donev (1989) and Cook and Nachtsheim (1989). The BLKL algorithm of Atkinson and Donev (1989) first computes an n-point starting design which is then improved by substituting a design point with a point from the list of candidate points until no further improvement in $\mathcal{D}$-efficiency can be made. The starting design is partly generated in a random fashion and completed by a greedy heuristic. In order to avoid being stuck in a locally optimal design, more than one starting design is generated and the exchange procedure is repeated. Each repetition of these steps is called a try. In contrast, Cook and Nachtsheim (1989) only use one try. In order to obtain a starting design, they compute a p-point design for model (4.9) and use these points to compose a nonsingular blocking design. The resulting design is improved by exchanging design points with candidate points and by interchanging design points from different blocks.

In the generic algorithm described here, more than one try is used and the starting designs are partly composed in a random fashion and completed by sequentially adding the candidate point with the largest prediction variance. In order to improve the initial design, both exchanging design points with candidate points and interchanging observations from different blocks are considered. The input to the algorithm consists of the number of observations n, the number of blocks b, the block sizes k_i $(i = 1, 2, \ldots, b)$, the number of model parameters p, the order of the model, the number of explanatory variables m and the structure of their polynomial expansion, and the number of tries t. In addition, an estimate of η must be provided. Typically, information on η is available from prior experiments of a similar kind. For example, Khuri (1992) analyzes an experiment in which the effect of temperature and time on shear strength is investigated and obtains $\hat{\eta} = 0.2928$, while Gilmour and Trinca (2000) analyze a pastry dough mixing experiment and obtain $\hat{\eta} = 10.01$ for one response and $\hat{\eta} = 1.40$ for another (see also Section 3.3.1). In the former experiment, the blocks were the batches of experimental material randomly selected from the warehouse supply. In the latter, the observations within each block were carried out on the same day. By default, the algorithm computes the grid of candidate points $G = \{1, 2, \ldots, g\}$ as in Atkinson and Donev (1992): the design region is assumed to be hypercubic and is taken as $[-1, +1]^k$. The grid points are chosen from the 2^m, 3^m, 4^m, ... factorial design depending on whether the model contains linear, quadratic, cubic or higher order terms. Alternatively, the user can specify G if another set of candidate points is desired. This is important when the design region is hyperspherical or when it is restricted. In order to find efficient designs, the set of candidate points should certainly include corner points and cover the entire design region. For example, the points of a 3^m factorial design are not sufficient to find an efficient design to estimate a quadratic model on a hyperspherical region. In that case, star points should be included in the set of candidate points. Finally, note that construction of a nonsingular design requires $n \geq p$ and $g \geq p$. Further details on the algorithm are given in Appendix A.

4.5.3 Algorithm evaluation

In order to evaluate the quality of our algorithm, we have used it 1,000 times in several situations for which the optimal design is known. For each of the problems investigated, we have used 8 values of the degree of correlation η: 0.1, 0.25, 0.5, 0.75, 1, 2, 5, 10. We have estimated the probability of finding the optimal design by dividing the number of times the optimum was found by the number of tries. It turns out that the probability of finding the optimal design is excellent. We will report average computation times for each design problem considered, even though computation time is in general not considered an issue in experimental design. As expected, the computation time is an increasing function of the number of observations n and the number of model parameters p. Moreover, the computation time increases with the degree of correlation η. The computer used was a 233MHz Pentium PC with 64MB RAM.

Design problem 1

Firstly, we have considered the optometry experiment described by Chasalow (1992). The purpose of the experiment was to estimate a quadratic regression model in one variable. The levels of the experimental variable were -1, 0 and +1. The design problem is to determine the optimal treatment levels and to assign them to blocks of size two. We have considered $b = 36, 48, 49$ and 60. For those cases where b is a multiple of three, the $\mathcal{D}$-optimal design has $b/3$ blocks with observations at -1 and 0, $b/3$ blocks with observations at -1 and +1 and $b/3$ blocks with observations at 0 and +1. For $b = 49$, one of these blocks is replicated 17 times, whereas the other two are replicated 16 times. It turns out that the optimal design is found in each individual try, independent of the number of blocks and the degree of correlation specified. The average computation time per try is around 0.200s when $n = 36$. When $n = 49$, it amounts to 0.460s when $\eta = 0.1$ and to 1.043s when $\eta = 10$. When $b = 60$, the average computation time per try is at most 1.856s.

Design problem 2

Next, we have used our algorithm to compute $\mathcal{D}$-optimal designs for the first example given in Section 4.3.1. The problem is to find a design with two blocks of four observations to estimate the linear effects and two-factor interactions of the three experimental variables. The estimated probability of finding the best design is equal to one.

Design problem 3

We have also investigated whether our algorithm was able to find an optimal design for the second example given in Section 4.3.1, that is a 32-point design for model (4.16) arranged in four blocks of four observations and two blocks of eight observations. An optimal design was given in Table 4.2. The probability of finding the optimum is equal to one in all cases. The average computation time per try increases monotonely from 1.079s when $\eta = 0.1$ to 1.488s when $\eta = 10$.

Design problem 4

We also used our algorithm to compute the optimal design for the mixture experiment described in Section 4.3.3. For the experiment, 10 blocks of 3 observations were available to estimate a second order Scheffé polynomial in three factors. For this design problem, the optimal design is found in each individual try. The average computation time ranges from 0.085s, when η is close to zero, to 0.101s, when η is as large as 10.

Design problem 5

Finally, we have investigated whether our algorithm was capable of finding the balanced incomplete block designs of Table 1.7 and Table 4.3 and the lattice design of Table 1.8. It is known that these designs are $\mathcal{D}$-optimal for any degree of correlation. The probability of finding the designs of Table 1.7 and Table 4.3 is one. The average computation time per try is small and does not exceed 0.092s, even when η is large. For the lattice design, the optimal design is found in more than 99.3% of the tries when $\eta \geq 0.25$. When $\eta = 0.1$, 98.6% of the tries leads to the optimal balanced incomplete block design. The average computation time per try lies between 0.328s and 0.389s.

4.6   Computational results

Firstly, we will concentrate on $\mathcal{D}$-optimal designs with three levels for each factor: -1, 0 and +1. Therefore, design points were chosen from the 3^m factorial design. Although these designs are often slightly less efficient than designs with more factor levels, they provide useful insights. Moreover, they are often used in practice because experimenters are reluctant to the use of many different factor levels. The design of experiments with more than three factor levels will receive attention in Section 4.8.

We have generated $\mathcal{D}$-optimal block designs for various combinations of the number of observations n, the number of blocks b and the number of experimental variables m. When $n > p + b$, we were able to compare the random block designs generated by the algorithm from the previous section to the fixed block designs generated by the algorithms of Atkinson and Donev (1989) and Cook and Nachtsheim (1989). Design points were chosen from the 3^m factorial design. It turns out that taking into account the random nature of the block effects, and thereby the compound symmetric

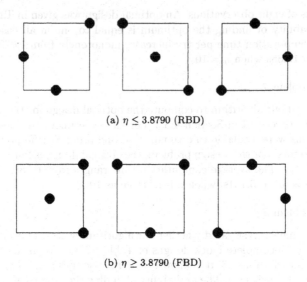

(a) $\eta \leq 3.8790$ (RBD)

(b) $\eta \geq 3.8790$ (FBD)

Figure 4.2: $\mathcal{D}$-optimal three-level designs with three blocks of size three for the full quadratic model in two variables.

error structure, is especially useful when the number of experimental variables exceeds two and when the model is not purely linear. Two design problems will illustrate our computational results.

Firstly, consider the 9-point $\mathcal{D}$-optimal design with three blocks of size three. The optimal designs for $\eta \leq 3.8790$ and $\eta \geq 3.8790$ are displayed in Figure 4.2 and are denoted by RBD and FBD respectively. The design for $\eta \geq 3.8790$, that is FBD, coincides with the $\mathcal{D}$-optimal fixed block design. An interesting feature of the design for small η is that its projection, obtained by ignoring the blocks, results in the 3^2 factorial, which is the $\mathcal{D}$-optimal design for the uncorrelated model (4.9). For the design in Figure 4.2b this is not the case.

In order to compare the $\mathcal{D}$-criterion values of random block designs and fixed block designs under different degrees of correlation, we have computed the relative $\mathcal{D}$-efficiencies

$$\left\{ \frac{|\mathbf{X}'\mathbf{V}^{-1}\mathbf{X}|}{|\mathbf{A}'\mathbf{V}^{-1}\mathbf{A}|} \right\}^{1/p}, \tag{4.24}$$

where $\mathbf{X}$ is the design matrix of the random block design under consideration and $\mathbf{A}$ is the design matrix of the fixed block design for the same design problem. In Figure 4.3, the relative efficiency of the random block

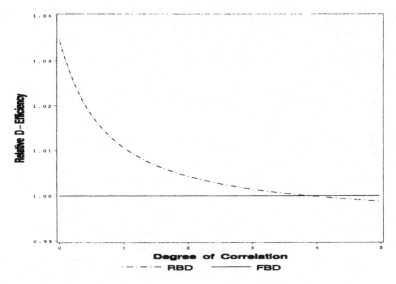

Figure 4.3: Comparison of the $\mathcal{D}$-efficiency of the designs in Figure 4.2 for the full quadratic model in two variables.

design in Figure 4.2a with respect to the fixed block design in Figure 4.2b is displayed. It is clear that RBD outperforms FBD for $\eta < 3.8790$. For η close to zero, the former is more than 3% more efficient than the latter. However, the efficiency gain obtained by taking into account the correlation in the design phase decreases as the degree of correlation increases. For $\eta \geq 3.8790$, FBD is better than RBD. This is consistent with the fact that the optimal random block design for large η is equal to the optimal fixed block design.

The picture for more complicated models looks somewhat different. Consider for example a full quadratic model in four variables and suppose six blocks of four observations are available for experimentation. For this design problem, we have found one random block design that is optimal for $\eta \leq 0.00108$ and one that is optimal for $0.00108 \leq \eta \leq 2685048.042$. Let both designs be denoted by RBD1 and RBD2 respectively. The projection of RBD1 obtained by ignoring the blocks yields the $\mathcal{D}$-optimal design for the uncorrelated full quadratic model in four variables, while the projection of RBD2 is slightly less efficient. The projection of the fixed block design for this design problem (FBD) is not even close to $\mathcal{D}$-optimal for the uncorrelated model. Both designs are compared to the fixed block design FBD for the same design problem in Figure 4.4. For $\eta \leq 0.12269$, RBD1 is better than FBD. However, FBD is outperformed by RBD2 for any practical value of η. Compared to FBD, the $\mathcal{D}$-efficiency is increased by more than

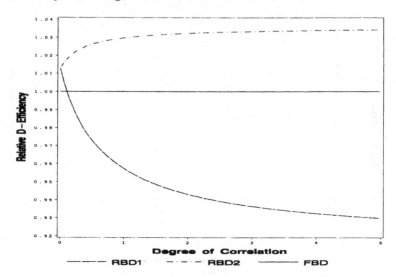

Figure 4.4: Comparison of the $\mathcal{D}$-efficiency of the random block designs RBD1 and RBD2 to the fixed block design FBD with six blocks of four observations for the full quadratic model in four variables.

1% for small degrees of correlation and by more than 3% when the degree of correlation exceeds unity.

We obtained similar results for both first and second order models for other combinations of n, b and k. In general, we can conclude that $\mathcal{D}$-optimal designs in the presence of random block effects differ from $\mathcal{D}$-optimal designs in the presence of fixed block effects. While the projection of the random block designs is in many cases $\mathcal{D}$-optimal for the uncorrelated model, this is not at all true for the projection of the fixed block design. Therefore, the construction of the random block designs can be seen as assigning observations of a highly efficient design for the uncorrelated model to blocks in order to obtain an efficient design for the correlated model. On the contrary, an efficient design in the presence of fixed block effects is obtained from an inefficient design for the uncorrelated model. Computational results also indicate that the random block designs are robust to misspecification. Typically, only one, two or three different random block designs were found for a given design problem. As a result, these designs are optimal for wide ranges of η. Precise prior knowledge of the degree of correlation η is therefore not needed to generate $\mathcal{D}$-optimal random block designs. It should be pointed out that sometimes, unlike the example given in Figure 4.2, the fixed block design turns out to be the optimal random block design as well for relatively small values of η (see, for example, the next section).

Table 4.4: Factor levels used in the pastry dough mixing experiment.

Moisture content	Screw speed	Flow rate
76%	300 rpm	30.0 kg/h
79%	350 rpm	37.5 kg/h
82%	400 rpm	45.0 kg/h

4.7 Pastry dough mixing experiment

The pastry dough mixing experiment, which was introduced in Section 3.3.1, is a nice example of an experimental situation in which a blocked experiment is needed. In the experiment, the effects of the factors moisture content, screw speed of the mixer, and flow rate on the color of a pastry were investigated. The response variables measured refer to the light reflectance in several bands of the spectrum. For the experiment, seven days were available, and four runs could be performed per day. A full quadratic model in the three explanatory factors was to be estimated. Therefore, three levels were used for each factor. These levels are displayed in Table 4.4. However, as is usually done in the design literature, we will represent the factor levels in coded form in the sequel of this section.

The set of treatments chosen by Trinca and Gilmour (2000) and Gilmour and Trinca (2000) for this experiment is a central composite design with the two-level factorial portion, that is the corner points of the cubic design region, duplicated and six center points. These treatments were assigned to the blocks according to Figure 4.5a. It can be seen that this design consists of two replicates of the corner points, one replicate at the midpoint of the faces of the cube, and six replications of the center runs. It is easy to verify that the average levels of x_1, x_2 and x_3 are zero so that the design is orthogonally blocked with respect to the main effect terms. This is not the case for the interactions and the quadratic effects. The projection of this design obtained by ignoring the blocks is symmetric. It is displayed in Figure 4.6a. The $\mathcal{D}$-optimal design for the pastry dough experiment is given in Figure 4.5b. This design turns out to be optimal for every value of η examined. It only contains two center points, and midpoints of the edges of the cube instead of midpoints of its faces. This can be verified by examining its projection in Figure 4.6. The projection is nearly symmetric and, although it is not orthogonal with respect to the main effect terms, the blocked design is almost orthogonal with respect to all model terms. This can be verified by computing the efficiency factor defined by John and Williams (1995), which is a measure for the orthogonality of a blocked design. The $\mathcal{D}$-optimal design has an efficiency factor of 96.10%, whereas the modified central composite design yields an efficiency factor of 91.90% only.

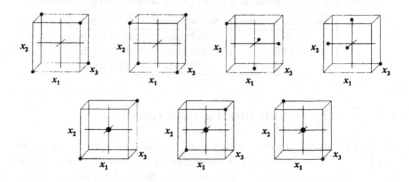

(a) Design used by Gilmour and Trinca (2000).

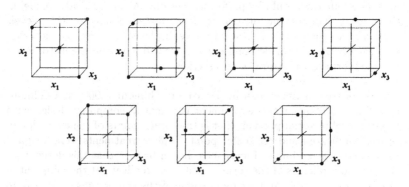

(b) $\mathcal{D}$-optimal design.

Figure 4.5: Design options for the pastry dough mixing experiment. • is a design point, ⊙ is a design point replicated twice.

It is no surprise that the $\mathcal{D}$-optimal design outperforms the modified central composite design in terms of $\mathcal{D}$-efficiency. However, it is also much better in terms of $\mathcal{A}$-efficiency and $\mathcal{V}$-efficiency. In Figure 4.7, the relative $\mathcal{D}$-, $\mathcal{A}$- and $\mathcal{V}$-efficiencies of the $\mathcal{D}$-optimal design from Figure 4.5b with respect to the modified central composite design from Figure 4.5a are displayed. The relative $\mathcal{D}$-efficiency increases with the degree of correlation η and is well above 115% unless η is close to zero. The relative $\mathcal{A}$- and $\mathcal{V}$-efficiencies decrease with η. They are substantially greater than 100% for small degrees of correlation and tend to 100% when η approaches 100.

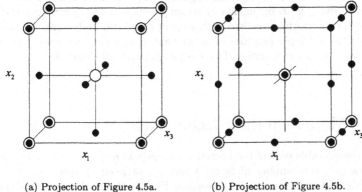

(a) Projection of Figure 4.5a.   (b) Projection of Figure 4.5b.

Figure 4.6: Projections of the designs from Figure 4.5 obtained by ignoring the blocks. • is a design point, ⊙ is a design point replicated twice, and ○ is a design point replicated six times.

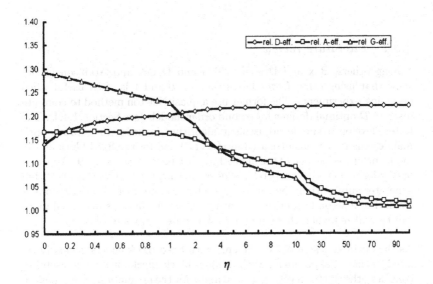

Figure 4.7: Relative $\mathcal{D}$-, $\mathcal{A}$- and $\mathcal{V}$-efficiencies of the $\mathcal{D}$-optimal design in Figure 4.5b with respect to the modified central composite design in Figure 4.5a.

The pastry dough mixing experiment is an excellent illustration of the usefulness of the optimal design approach to construct blocked industrial experiments. The $\mathcal{D}$-optimality criterion produces a nearly orthogonal design which performs excellent with respect to several design criteria. As a result, this example provides evidence that it may not be a good idea to modify standard response surface designs in practical situations.

4.8 More than three factor levels

In a considerable part of the literature on optimal response surface designs in practice, the number of factor levels is restricted to three: -1, 0 and 1. This is due to a number of reasons. Firstly, the levels of the optimal continuous design for second order models with uncorrelated observations are -1, 0 and 1, as is demonstrated by Farrell, Kiefer and Walbran (1967). Secondly, popular standard response surface designs have only a limited number of factor levels. Thirdly, experimenters prefer using only a small number of factor levels. In many instances, however, it is worthwhile to consider other factor levels as well.

4.8.1 Literature review

Among others, Box and Draper (1971) and Donev and Atkinson (1988) show that using other factor levels than -1, 0 and 1 is often useful when designing experiments. The former use a quasi-Newton method to compute discrete $\mathcal{D}$-optimal designs for second order models (see Section 1.4.4). The latter develop a simple adjustment algorithm in order to improve optimal designs for the uncorrelated model (1.3) and for the fixed block effects model produced by one of the algorithms discussed in Section 1.9. For a second order model with uncorrelated observations, the adjustment algorithm generates substantially better designs when the number of observations is small. For a second order model with fixed block effects, substantial gains can be realized when there are only a few observations per block.

As the optimal design of experiments with fixed block effects is closely related to that of experiments with random block effects, it may be useful to consider other factor levels than -1, 0 and 1 for the estimation of the random block effects model (4.2) as well. The results of Atkins (1994), Cheng (1995) and Atkins and Cheng (1999) on $\mathcal{D}$-optimal continuous designs for the random block effects model also point in this direction. Cheng (1995) and Atkins and Cheng (1999) use an approximate theory to compute $\mathcal{D}$-optimal continuous designs for quadratic regression on $[-1, 1]$. They point out that the weights[1] of the different blocks as well as the factor levels in the $\mathcal{D}$-optimal design depend on η. For instance, Cheng (1995) shows that the continuous $\mathcal{D}$-optimal design with blocks of size two is supported on the blocks $(1; -a_\eta)$, $(-1; a_\eta)$ and $(-1; 1)$ $(a_\eta \geq 0)$ with weights $\varepsilon_\eta/2$, $\varepsilon_\eta/2$ and $1 - \varepsilon_\eta$ respectively. For example, for $\eta = 0.25$ the optimal values for a_η and ε_η amount to 0.059255 and 0.675536 respectively. Cheng (1995) shows that $a_\eta \rightarrow 0$ and $\varepsilon_\eta \rightarrow 2/3$ when η approaches zero. Atkins (1994) uses the same approximate theory to compute $\mathcal{D}$-optimal designs for general design problems in the presence of random block effects. Although these continuous designs are of little use in practice, especially when the number of blocks is small, they provide at least one important insight: other factor levels than -1, 0 and 1 occur in the $\mathcal{D}$-optimal continuous designs. In the sequel of this section, we will show that this is also the case for $\mathcal{D}$-optimal discrete designs for the random block effects model. For this purpose, we will first investigate whether using a finer grid of candidate points enables us to construct more efficient designs. Next, we will develop an adjustment algorithm, and finally, we combine the search over a finer grid and the adjustment algorithm.

In total, we have investigated 28 different combinations of the number of experimental variables m, the number of observations n, the number of blocks b and the block sizes k_i $(i = 1, 2, \ldots, b)$. The left panel of Table 4.5 contains the key parameters of the design problems considered. Although we have investigated 10 different values of the degree of correlation η, only the results for $\eta = 0.01$, 1 and 10 will be reported. Consequently, the computational results for $28 \times 3 = 84$ design problems will be given. In the sequel of this section, we will briefly discuss the three different approaches for finding better designs than those obtained by using the coarse 3^m grid. Next, we will identify those cases where the largest improvements can be realized. Finally, we take a closer look at the $\mathcal{D}$-optimal designs for a small design problem.

4.8.2   Finer grid

As a start, we have compared the $\mathcal{D}$-optimal designs for a full second order model obtained by using an 11^m grid on the experimental region $[-1, 1]^m$ to the designs obtained by using the coarse 3^m grid. In many instances, using the finer grid yields substantially better designs. The improvements in $\mathcal{D}$-efficiency for the 84 design problems are displayed in the second panel of Table 4.5. For each design problem, 1,000 tries were used.

---

[1] While a continuous design for an unblocked experiment is represented by a measure on the set of design points, a continuous design for a blocked experiment is represented by a measure on the set of blocks in the experiment.

4.8.3   Adjustment algorithm

We have also modified the adjustment algorithm of Donev and Atkinson (1988) for the design problem considered here. As a result of applying the design construction algorithm described in Section 4.5.2, a locally optimal design is obtained. The adjustment algorithm is a method of determining this local optimum more precisely. It calculates the effect on the $\mathcal{D}$-optimality criterion of moving each design point a small amount, called a step, along each factor axis. The change that generates the greatest increase is carried out and the process is repeated until no further progress can be made. If no improvement can be found, the step length is halved and the process is repeated. The algorithm stops when the step length becomes smaller than a prespecified minimum step length. At most $2mn$ design changes are evaluated in each stage. When one or more design points lie on the boundary of the experimental region, the number of changes evaluated is less than $2mn$ because points outside the experimental region are omitted. The input to the adjustment algorithm includes the initial step length and the minimum step length. A formal description of the algorithm is given in Appendix B.

As a starting design to the adjustment algorithm, we used the $\mathcal{D}$-optimal designs obtained from a search with 1,000 tries over the 3^m grid on the experimental region $[-1, 1]^m$. In some cases, the adjustment algorithm was unable to improve the starting design. A suboptimal design was then used as a starting design for the adjustment algorithm. The details of the step length reduction do not seem to influence the results. The improvements obtained by using the adjustment algorithm are displayed in Table 4.5. In many cases, the resulting designs were slightly better than those obtained by searching over the 11^m grid. However, the adjustment algorithm was unable to improve the local optimum when $m = 1$, $k_1 = k_2 = 3$, $k_3 = 2$ and $\eta = 10$, even though the search over the 11^m grid had shown that an improvement of at least 0.25% was possible. In quite a number of other cases, the adjustment algorithm was clearly inferior to the use of a finer grid. For example, the adjustment algorithm only found a design that was 0.31% better than the starting design when $m = 1$, $k_1 = 3$ $k_2 = k_3 = 2$ and $\eta = 10$, even though the search over a finer grid produced a design that was 2.76% more $\mathcal{D}$-efficient. This is because only one factor level is changed at a time.

4.8.4   Combined approach

It is clear that, in most cases, both the search over a finer grid and the adjustment algorithm are able to produce more efficient designs than a search over a coarse grid. Nevertheless, the adjustment algorithm is sometimes unable to improve the starting design due to the fact that only one

Table 4.5: Improvement in $\mathcal{D}$-efficiency obtained by using a finer grid, an adjustment algorithm and a combined approach.

	Design Problem				Finer grid			Adjustment			Combined approach		
m	k_1	k_2	k_3	n	$\eta=0.01$	$\eta=1$	$\eta=10$	$\eta=0.01$	$\eta=1$	$\eta=10$	$\eta=0.01$	$\eta=1$	$\eta=10$
1	2	2		4	0.00%	4.55%	10.21%	0.00%	4.90%	10.71%	0.00%	4.90%	10.71%
1	2	3		5	0.00%	0.23%	0.93%	0.00%	0.65%	1.26%	0.00%	0.65%	1.26%
1	2	2	2	6	0.00%	0.33%	1.62%	0.00%	0.78%	1.73%	0.00%	0.78%	1.73%
1	3	2	2	7	0.00%	1.41%	2.76%	0.00%	0.15%	0.31%	0.00%	4.53%	3.01%
1	3	3	2	8	0.00%	0.08%	0.25%	0.00%	0.17%	0.00%	0.00%	0.34%	0.70%
2	3	4		7	0.00%	0.30%	0.42%	0.36%	0.44%	0.51%	0.36%	0.44%	0.51%
2	4	4		8	0.28%	0.36%	0.40%	0.41%	0.12%	0.19%	0.41%	0.42%	0.46%
2	4	5		9	0.00%	0.00%	0.05%	0.00%	0.19%	0.28%	0.00%	0.00%	0.28%
2	2	8		10	0.00%	0.34%	0.46%	0.18%	0.48%	0.63%	0.18%	0.48%	0.63%
2	3	7		10	0.01%	0.35%	0.43%	0.19%	0.50%	0.59%	0.19%	0.50%	0.59%
2	4	6		10	0.00%	0.03%	0.09%	0.18%	0.31%	0.35%	0.18%	0.31%	0.35%
2	5	5		10	0.01%	0.47%	0.77%	0.19%	0.47%	0.38%	0.19%	0.50%	0.80%
2	5	6		11	0.06%	0.00%	0.01%	0.22%	0.26%	0.26%	0.23%	0.26%	0.26%
2	5	7		12	0.02%	0.14%	0.16%	0.19%	0.17%	0.17%	0.19%	0.24%	0.25%
2	2	11		13	0.00%	0.00%	0.02%	0.00%	0.13%	0.23%	0.00%	0.13%	0.23%
2	3	10		13	0.00%	0.00%	0.00%	0.00%	0.08%	0.11%	0.00%	0.08%	0.11%
2	4	9		13	0.00%	0.00%	0.00%	0.00%	0.04%	0.06%	0.00%	0.04%	0.06%
2	5	8		13	0.00%	0.00%	0.00%	0.00%	0.00%	0.00%	0.00%	0.00%	0.00%
2	6	7		13	0.00%	0.00%	0.00%	0.00%	0.04%	0.03%	0.00%	0.02%	0.05%
2	6	8		14	0.00%	0.00%	0.00%	0.00%	0.00%	0.00%	0.00%	0.00%	0.00%
2	7	8		15	0.00%	0.00%	0.00%	0.02%	0.01%	0.01%	0.02%	0.01%	0.01%
2	8	10		18	0.00%	0.00%	0.00%	0.00%	0.00%	0.00%	0.00%	0.00%	0.00%
2	2	3	3	8	0.30%	1.33%	2.23%	0.42%	1.45%	2.30%	0.43%	1.45%	2.30%
2	3	3	3	9	0.00%	0.81%	1.93%	0.00%	1.07%	1.49%	0.00%	1.07%	2.12%
2	3	3	4	10	0.01%	0.51%	0.69%	0.18%	0.66%	0.93%	0.18%	0.66%	0.93%
2	3	3	5	11	0.05%	0.46%	0.61%	0.22%	0.65%	0.76%	0.23%	0.65%	0.76%
2	3	4	5	12	0.03%	0.27%	0.32%	0.19%	0.42%	0.49%	0.19%	0.42%	0.49%
2	5	5	5	15	0.00%	0.00%	0.00%	0.02%	0.07%	0.07%	0.02%	0.07%	0.07%

factor level is changed at a time. Using a fine grid on the experimental region does not suffer from this restriction because each design point can be replaced by a point in which more than one factor level is modified. We have combined the methods described in Sections 4.8.2 and 4.8.3 in an attempt to improve the designs obtained by one of both methods separately. The results are displayed in Table 4.5. In 82 out of the 84 cases reported, the combined approach yielded the best result. For 69 of these design problems, the same result was obtained by simply applying the adjustment algorithm to the best three-level design. In 13 cases, the combined approach was strictly the best. The most striking improvements were realized when there is one experimental variable, $n = 4$ ($k_1 = k_2 = 2$) or $n = 7$ ($k_1 = 3, k_2 = k_3 = 2$), and $\eta = 1$ or 10. The combined approach was inferior to the use of the adjustment algorithm for the 2 remaining design problems considered.

4.8.5   Opportunities for large improvements

The improvement in $\mathcal{D}$-efficiency over the three-level $\mathcal{D}$-optimal designs largely depends on the parameters of the design problem. The smaller the number of observations available and the smaller the block sizes, the larger will be the potential increase in $\mathcal{D}$-efficiency. For example, it can be seen from Table 4.5 that the improvement when $m = 2$, $n = 15$, $k_1 = k_2 = k_3 = 5$ and $\eta = 1$ is nearly zero, whereas it amounts to 1.07% when $n = 9$, $k_1 = k_2 = k_3 = 3$. When $n = 9$, $k_1 = 4$, $k_2 = 5$ and $\eta = 10$, the improvement amounts to 0.28%, which is considerably worse than the 2.12% improvement that can be realized when $k_1 = k_2 = k_3 = 3$. In addition, the improvement in $\mathcal{D}$-efficiency tends to be larger for higher degrees of correlation η. For example, the improvement increases from 0.36% when $\eta = 0.01$ to 0.51% when $\eta = 10$ for the design problem with parameter values $n = 7$, $k_1 = 3$ and $k_2 = 4$.

4.8.6   An illustration

Consider again a blocked experiment with three blocks of size three for the estimation of a full quadratic model in two explanatory variables. The $\mathcal{D}$-optimal three-level designs for this problem are given in Figure 4.2. As indicated in Table 4.5, the combined adjustment approach produced substantially better designs. By means of a white circle, we have displayed the points of the $\mathcal{D}$-optimal designs for five different values of η in Figure 4.8. In order to visualize the differences between these designs and the $\mathcal{D}$-optimal three-level designs, we have displayed the latter by means of a black bullet. For the panels a to d, we have used the points of the three-level design in Figure 4.2a, while for panel e, we have used the points given in Figure 4.2b. In each panel of Figure 4.8, some of the bullets are invisible because the corresponding design points do not move as η increases. It is clear that the larger η, the more the design points move away from the three-level designs. This is consistent with the fact that the adjustment algorithms produce the largest improvements when η is large. The $\mathcal{D}$-optimal design for $\eta = 10$ turns out to be very efficient when the block effects are fixed. The corresponding $\mathcal{D}$-efficiency is 0.9197, which is more than the $\mathcal{D}$-efficiency of 0.9142 of the design given in Figure 13.3 on page 149 of Atkinson and Donev (1992). Probably, the reason why Atkinson and Donev (1992) got stuck in a suboptimal design is that the initial step size they used in their adjustment algorithm was too small. Finally, note that the three-level design represented by the black bullets in Figure 4.8d is not the $\mathcal{D}$-optimal three-level design for $\eta = 5$ because using it as an input to the adjustment algorithm leads to a local optimum. Instead, the $\mathcal{D}$-optimal three-level design for small values of η was displayed.

4.9   Efficiency of blocking

The efficiency of blocking cannot be expressed as a simple ratio of variance components when the block effects are random. Consider the orthogonally blocked central composite design of Table 2.1. When $\sigma_\epsilon^2 = \sigma_\gamma^2 = 0.5$, the variance-covariance matrix of the parameter estimate $\hat{\beta}$ is given by

$$\begin{bmatrix} 0.0375\,\mathbf{I}_3 & \mathbf{0}'_{3\times3} & \mathbf{0}_3 & \mathbf{0}_3 & \mathbf{0}_3 \\ \mathbf{0}'_{3\times3} & 0.0625\,\mathbf{I}_3 & \mathbf{0}_3 & \mathbf{0}_3 & \mathbf{0}_3 \\ \mathbf{0}'_3 & \mathbf{0}'_3 & 0.0379 & 0.0027 & 0.0027 \\ \mathbf{0}'_3 & \mathbf{0}'_3 & 0.0027 & 0.0379 & 0.0027 \\ \mathbf{0}'_3 & \mathbf{0}'_3 & 0.0027 & 0.0027 & 0.0379 \end{bmatrix}. \qquad (4.25)$$

When the central composite design is used in a completely randomized experiment with $\sigma^2 = 1$, the variance-covariance matrix is given by

$$\begin{bmatrix} 0.075\,\mathbf{I}_3 & \mathbf{0}'_{3\times3} & \mathbf{0}_3 & \mathbf{0}_3 & \mathbf{0}_3 \\ \mathbf{0}'_{3\times3} & 0.125\,\mathbf{I}_3 & \mathbf{0}_3 & \mathbf{0}_3 & \mathbf{0}_3 \\ \mathbf{0}'_3 & \mathbf{0}'_3 & 0.076 & 0.005 & 0.005 \\ \mathbf{0}'_3 & \mathbf{0}'_3 & 0.005 & 0.076 & 0.005 \\ \mathbf{0}'_3 & \mathbf{0}'_3 & 0.005 & 0.005 & 0.076 \end{bmatrix}. \qquad (4.26)$$

The determinants of these matrices are 3.5251E-10 and 6.8849E-13, so that the relative $\mathcal{D}$-efficiency of blocking with respect to not blocking amounts to 1.8661. This value is not equal to the ratio of σ^2 to σ_ϵ^2, which is 2 in this example.

4.10   Optimal number of blocks and block sizes

Frequently, the number of blocks and the block size are dictated by the experimental situation. For instance, in the optometry experiment, the block

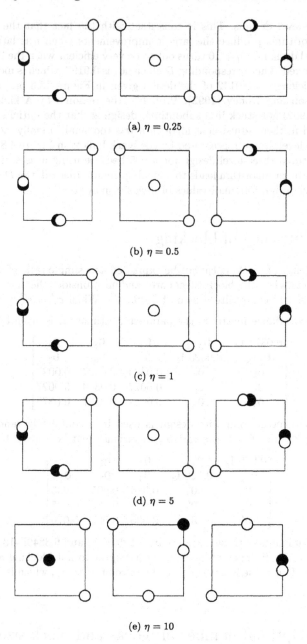

Figure 4.8: $\mathcal{D}$-optimal designs with three blocks of size three for the full quadratic model in two variables. The adjusted design points are represented by a ∘. The points of the $\mathcal{D}$-optimal three-level design are represented by a •.

Table 4.6: $\mathcal{D}$-criterion values of the 24-point $\mathcal{D}$-optimal random block designs for different numbers of blocks and for different values of η.

		η					
b	k	0.1	0.3	0.5	1	2	4
2	12	1.216E16	7.051E16	3.951E17	1.587E19	2.960E21	3.900E24
3	8	1.410E16	8.294E16	5.015E17	1.912E19	4.408E21	4.883E24
4	6	1.485E16	9.182E16	5.198E17	2.097E19	4.803E21	5.382E24
6	4	1.339E16	6.872E16	3.551E17	1.337E19	2.901E21	3.018E24
8	3	1.155E16	4.614E16	1.925E17	4.793E18	7.946E20	6.649E23
12	2	9.176E15	2.204E16	5.891E16	6.220E17	3.213E19	7.082E21

size cannot exceed two since each human has only two eyes. In other cases, it may well be possible that the experimenter is at liberty to determine the number of blocks and the block sizes given the number of observations. For example, an experiment with 24 observations could be carried out by two researchers, performing twelve observations each, or by three researchers, performing only eight observations each, and so on. Assume that the 24-point experiment is to be designed for the estimation of the full second order model in four variables and that the experimenter has the choice between two blocks of twelve observations ($b = 2$, $k = 12$), three blocks of eight observations ($b = 3$, $k = 8$), four blocks of six observations ($b = 4$, $k = 6$), six blocks of four observations ($b = 6$, $k = 4$), eight blocks of three observations ($b = 8$, $k = 3$) and twelve blocks of two observations ($b = 12$, $k = 2$). For each of these experimental settings, we have computed $\mathcal{D}$-optimal random block designs for different values of η and have displayed their $\mathcal{D}$-criterion values in Table 4.6. The $\mathcal{D}$-criterion values were calculated holding $\sigma_\varepsilon^2 + \sigma_\gamma^2 = 1$. It turns out that, for any η, carrying out an experiment with four blocks of six observations is the best design option. The second best option is the experiment with three blocks. The worst choices are the designs with eight and twelve blocks.

Two other remarks should be made at this point. Firstly, a $\mathcal{D}$-optimal 24-point design for the uncorrelated model (4.9) with $\sigma_\varepsilon^2 = 1$ has a $\mathcal{D}$-criterion value of 6.577E15. As a result, it is less efficient than every block design considered in Table 4.6. Secondly, designs with a homogeneous block size are not necessarily the best design option. Atkinson and Donev (1989) already obtained this result for fixed block effects. We have found similar results when the blocks are random. For instance, the 10-point $\mathcal{D}$-optimal random block design with one block of four observations and one block of six observations has a better $\mathcal{D}$-criterion value than the $\mathcal{D}$-optimal random block design with two blocks of five observations. The $\mathcal{D}$-optimal designs for both situations are displayed in Figure 4.9. They are optimal for any value of η and equivalent to the $\mathcal{D}$-optimal fixed block designs.

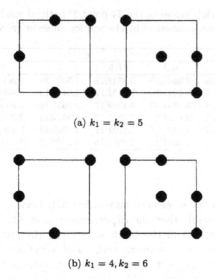

(a) $k_1 = k_2 = 5$

(b) $k_1 = 4, k_2 = 6$

Figure 4.9: 10-point $\mathcal{D}$-optimal designs with two blocks.

4.11   Optimality of orthogonal blocking

In Section 4.3.1, we already showed that orthogonally blocking first order designs is an optimal strategy for a given design matrix $\mathbf{X}$. In this section, we show that orthogonal blocking is an optimal blocking strategy for models of any order provided the block size is homogeneous. Substituting the condition for orthogonality (2.39) in (4.7) yields the information matrix of an orthogonally blocked experiment:

$$
\begin{aligned}
\mathbf{M}_{\text{orth.}} &= \frac{1}{\sigma_\varepsilon^2}\{\mathbf{X}'\mathbf{X} - \sum_{i=1}^{b} \frac{\eta}{1 + k_i\eta}(\frac{k_i}{n}\mathbf{X}'\mathbf{1}_n)(\frac{k_i}{n}\mathbf{X}'\mathbf{1}_n)'\}, \\
&= \frac{1}{\sigma_\varepsilon^2}\{\mathbf{X}'\mathbf{X} - \sum_{i=1}^{b} \frac{k_i^2\eta}{n^2(1 + k_i\eta)}(\mathbf{X}'\mathbf{1}_n)(\mathbf{X}'\mathbf{1}_n)'\}.
\end{aligned}
\tag{4.27}
$$

Arranging the n observations so that condition (2.39) is not satisfied yields a design that is not orthogonally blocked. In that case,

$$
\mathbf{X}_i'\mathbf{1}_{k_i} = \frac{k_i}{n}\mathbf{X}'\mathbf{1}_n + \boldsymbol{\delta}_i, \qquad (i = 1, 2, \ldots, b),
\tag{4.28}
$$

where at least one $\delta_i \neq 0_p$. Summing (4.28) for all $i = 1, 2, \ldots, b$ gives us

$$\sum_{i=1}^{b} \mathbf{X}'_i 1_{k_i} = \sum_{i=1}^{b} (\frac{k_i}{n} \mathbf{X}' 1_n + \delta_i),$$

$$\mathbf{X}' 1_n = \sum_{i=1}^{b} \frac{k_i}{n} \mathbf{X}' 1_n + \sum_{i=1}^{b} \delta_i,$$

$$\mathbf{X}' 1_n = \frac{1}{n} \mathbf{X}' 1_n \sum_{i=1}^{b} k_i + \sum_{i=1}^{b} \delta_i,$$

$$\mathbf{X}' 1_n = \frac{n}{n} \mathbf{X}' 1_n + \sum_{i=1}^{b} \delta_i,$$

so that

$$\sum_{i=1}^{b} \delta_i = 0_p.$$

The information matrix on $\boldsymbol{\beta}$ can then be written as

$$
\begin{aligned}
\mathbf{M}_{\text{n.orth.}} &= \frac{1}{\sigma_\varepsilon^2} \{ \mathbf{X}'\mathbf{X} - \sum_{i=1}^{b} \frac{\eta}{1 + k_i \eta} (\frac{k_i}{n} \mathbf{X}' 1_n + \delta_i)(\frac{k_i}{n} \mathbf{X}' 1_n + \delta_i)' \}, \\
&= \mathbf{M}_{\text{orth.}} - \boldsymbol{\Delta}'_1 \boldsymbol{\Delta}_1 \qquad\qquad (4.29) \\
&\quad - \frac{1}{\sigma_\varepsilon^2} \{ \sum_{i=1}^{b} \frac{k_i \eta}{n(1 + k_i \eta)} (\mathbf{X}' 1_n) \delta'_i + \sum_{i=1}^{b} \frac{k_i \eta}{n(1 + k_i \eta)} \delta_i (\mathbf{X}' 1_n)' \},
\end{aligned}
$$

where

$$\boldsymbol{\Delta}_1 = \frac{\sqrt{\eta}}{\sigma_\varepsilon} \left[\frac{\delta_1}{\sqrt{1 + k_1 \eta}} \quad \frac{\delta_2}{\sqrt{1 + k_2 \eta}} \quad \cdots \quad \frac{\delta_b}{\sqrt{1 + k_b \eta}} \right]'.$$

When the block size is homogeneous and equal to k, (4.29) becomes

$$
\begin{aligned}
\mathbf{M}_{\text{n.orth.}} &= \mathbf{M}_{\text{orth.}} - \boldsymbol{\Delta}'_1 \boldsymbol{\Delta}_1 \\
&\quad - \frac{1}{\sigma_\varepsilon^2} \{ \sum_{i=1}^{b} \frac{k \eta}{n(1 + k \eta)} (\mathbf{X}' 1_n) \delta'_i + \sum_{i=1}^{b} \frac{k \eta}{n(1 + k \eta)} \delta_i (\mathbf{X}' 1_n)' \}, \\
&= \mathbf{M}_{\text{orth.}} - \boldsymbol{\Delta}'_1 \boldsymbol{\Delta}_1 \qquad\qquad (4.30) \\
&\quad - \frac{k \eta}{n(1 + k \eta)\sigma_\varepsilon^2} \{ (\mathbf{X}' 1_n)(\sum_{i=1}^{b} \delta'_i) + (\sum_{i=1}^{b} \delta_i)(\mathbf{X}' 1_n)' \}, \\
&= \mathbf{M}_{\text{orth.}} - \boldsymbol{\Delta}'_1 \boldsymbol{\Delta}_1.
\end{aligned}
$$

As a result, the difference

$$\mathbf{M}_{\text{orth.}} - \mathbf{M}_{\text{n.orth.}} = \boldsymbol{\Delta}'_1 \boldsymbol{\Delta}_1 \qquad\qquad (4.31)$$

is nonnegative definite. Therefore, when the block size is homogeneous, orthogonally blocked designs will be better than designs that are not blocked orthogonally with respect to any generalized optimality criterion, e.g. the $\mathcal{D}$- and $\mathcal{A}$-optimality criteria, for any given $\mathbf{X}$.

Whether orthogonality is a guarantee for optimality when the block size is heterogeneous remains an open question. When $\eta \to 0$, $k_i\eta/n(1 + k_i\eta) \to 0$ and the information matrix (4.29) reduces to (4.30). When $\eta \to +\infty$, $k_i\eta/n(1 + k_i\eta) \to 1/n$. In that case, the information matrix (4.29) also simplifies to (4.30). As a result, orthogonal blocking is optimal in these extreme cases, even when the block size is heterogeneous. Orthogonal blocking also turns out to be optimal when the number of experimental observations is large. As a matter of fact, when $n \to +\infty$, $k_i\eta/n(1 + k_i\eta) \to 0$.

Appendix A. Design construction algorithm

We denote the set of g candidate points by G, the set of b blocks by B, the set of k_i not necessarily distinct design points belonging to the ith block of a given design D by D_i $(i = 1, 2, \ldots, b)$ and the $\mathcal{D}$-criterion value of a given design D by $\mathcal{D}$. The best design found at a given time by the algorithm will be denoted by D^*. Its blocks will be denoted by D_i^* $(i = 1, 2, \ldots, b)$ and the corresponding $\mathcal{D}$-criterion value by $\mathcal{D}^*$. For simplicity, we denote the information matrix of the experiment by $\mathbf{M}$. The singularity while constructing a starting design is overcome by using $\mathbf{M} + \omega\mathbf{I}$ instead of $\mathbf{M}$ with ω a small positive number. Finally, we denote the number of tries by t and the number of the current try by t_c. The algorithm starts by specifying the set of grid points $G = \{1, 2, \ldots, g\}$ and proceeds as follows:

1. Set $\mathcal{D}^* = 0$ and $t_c = 1$.

2. Set $\mathbf{M} = \omega\mathbf{I}$ and $D_i = \emptyset$ $(i = 1, 2, \ldots, b)$.

3. Generate starting design.
   - (a) Randomly choose r $(1 \leq r \leq p)$.
   - (b) Do r times:
       - i. Randomly choose $i \in G$.
       - ii. Randomly choose $j \in B$.
       - iii. If $\#D_j < k_j$, then $D_j = D_j \cup \{i\}$, else go to step ii.
       - iv. Update $\mathbf{M}$.
   - (c) Do $n - r$ times:
       - i. Determine $i \in G$ with largest prediction variance.
       - ii. Randomly choose $j \in B$.
       - iii. If $\#D_j < k_j$, then $D_j = D_j \cup \{i\}$, else go to step ii.
       - iv. Update $\mathbf{M}$.

4. Compute $\mathbf{M}$ and $\mathcal{D}$. If $\mathcal{D} = 0$, then go to step 2, else continue.

5. Set $\kappa = 0$.

6. Evaluate exchanges.
   (a) Set $\delta = 1$.
   (b) $\forall i \in B, \forall j \in D_i, \forall k \in G, j \neq k$:
       i. Compute the effect $\delta_{ij}^k = \mathcal{D}'/\mathcal{D}$ of exchanging j by k in the ith block.
       ii. If $\delta_{ij}^k > \delta$, then $\delta = \delta_{ij}^k$ and store i, j and k.

7. If $\delta > 1$, then go to step 8, else go to step 9.

8. Carry out best exchange.
   (a) $D_i = D_i \backslash \{j\} \cup \{k\}$.
   (b) Update $\mathbf{M}$ and $\mathcal{D}$.
   (c) Set $\kappa = 1$.

9. Evaluate interchanges.
   (a) Set $\delta = 1$.
   (b) $\forall i, j \in B, i < j, \forall k \in D_i, \forall l \in D_j, k \neq l$:
       i. Compute the effect $\delta_{ik}^{jl} = \mathcal{D}'/\mathcal{D}$ of moving k from block i to j and l from block j to i.
       ii. If $\delta_{ik}^{jl} > \delta$, then $\delta = \delta_{ik}^{jl}$ and store i, j, k and l.

10. If $\delta > 1$, then go to step 11, else go to step 12.

11. Carry out best interchange.
    (a) $D_i = D_i \backslash \{k\} \cup \{l\}$.
    (b) $D_j = D_j \backslash \{l\} \cup \{k\}$.
    (c) Update $\mathbf{M}$ and $\mathcal{D}$.
    (d) Set $\kappa = 1$.

12. If $\kappa = 1$, then go to step 5, else go to step 13.

13. If $\mathcal{D} > \mathcal{D}^*$, then $\mathcal{D}^* = \mathcal{D}$, $\forall i \in B : D_i^* = D_i$.

14. If $t_c < t$, then $t_c = t_c + 1$ and go to step 2, else stop.

Appendix B. Adjustment algorithm

We denote by s the step length and by S the minimum step length. The other notation used in the formal description of the adjustment algorithm given here is identical to that used in Appendix A. Let the starting design $D = \{1, 2, \ldots, n\}$ be composed of n design points with coordinates $\mathbf{c}_i = (c_{i1}, c_{i2}, \ldots, c_{im})$, $i = 1, 2, \ldots, n$, let J be the set of all integers up to m and let K be the set of the integers 1 and 2. The steps of the adjustment algorithm are as follows:

1. Specify s and S.

2. Compute the determinant $\mathcal{D}$ and the information matrix $\mathbf{M}$ of the starting design.

3. Evaluate design changes.
   (a) Set $\delta = 1$.
   (b) $\forall i \in D, \forall j \in J, \forall k \in K$:
       i. Compute the effect $\delta_{ijk} = \mathcal{D}'/\mathcal{D}$ of replacing the jth coordinate of the ith design point c_{ij} with $c_{ij} + s \times (-1)^k$.
       ii. If $\delta_{ijk} > \delta$, then $\delta = \delta_{ijk}$ and store $i^* = i$, $j^* = j$ and $k^* = k$.

4. If $\delta > 1$, then go to step 5, else go to step 6.

5. Carry out the best exchange.
   (a) Replace $c_{i^* j^*}$ with $c_{i^* j^*} + s \times (-1)^{k^*}$.
   (b) Update $\mathcal{D}$ and $\mathbf{M}$ and go to step 3.

6. Set $s = s/2$.

7. If $s \geq S$, go to step 3, else stop.

5

Optimal Designs for Quadratic Regression on One Variable and Blocks of Size Two

In this chapter, exact $\mathcal{D}$-optimal designs are derived for an optometry experiment for the estimation of a quadratic polynomial in one explanatory variable. Two observations are made for each subject participating in the experiment, so that each subject serves as a block of two possibly correlated observations. The exact $\mathcal{D}$-optimal designs for this problem are compared to the best possible three-level designs and to the continuous $\mathcal{D}$-optimal designs.

5.1 Introduction

The purpose of this chapter is twofold. Firstly, it provides the reader with a series of exact $\mathcal{D}$-optimal designs for an optometry experiment with blocks of size two for the estimation of a quadratic model in one explanatory variable. It turns out that the designs presented here are substantially more efficient than the three-level designs proposed by Chasalow (1992). Secondly, the chapter provides the reader with a couple of interesting insights in the optimal design of experiments with correlated observations. It does not only demonstrate how the optimal designs depend on the extent to which the observations are correlated, but it also illustrates how the exact $\mathcal{D}$-optimal designs evolve towards the continuous $\mathcal{D}$-optimal designs derived by Cheng (1995) and Atkins and Cheng (1999) when the number of subjects available becomes large. In the next section, we give a concise description of the optometry experiment. The statistical model is introduced and the design criterion is derived in Section 5.3. The continuous $\mathcal{D}$-optimal designs are described in Section 5.4. In Section 5.5.1, we examine the $\mathcal{D}$-optimal three-level designs for the optometry experiment obtained by Chasalow (1992). Finally, we derive exact $\mathcal{D}$-optimal designs for several numbers of subjects in Section 5.5.2.

5.2   Optometry experiment

Chasalow (1992) describes an optometry experiment to investigate the health impact of wearing contact lenses. One consequence of wearing contact lenses is that the corneas, which are the clear structures that cover the front parts of the eyes including the irises and the pupils (see Figure 5.1), are exposed to a decreased level of O_2. The decrease in O_2 leads to the production of a weak acid and an increased flow of water into the cornea. The cornea has active mechanisms for regulating the in- and outflow of water in order to counteract the effect of the decreased O_2 level and to avoid damage from excess swelling or dessication. The eye's ability to regulate the water content of the cornea is usually referred to as corneal hydration control and naturally tends to decrease with age. However, it turns out that people who have worn contact lenses for some time tend to have corneas that look like much older people, at least with respect to corneal hydration control. In the optometry experiment, the effect of wearing contact lenses was imitated by exposing the human subject's eyes to a CO_2 treatment. Once it has passed the tear film, CO_2 mixes with the aqueous component of the tears to form a weak acid and activates the water regulating mechanism of the cornea. The CO_2 treatments were applied through a goggle covering the subject's eyes.

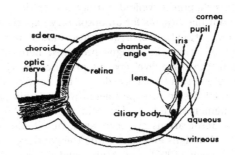

Figure 5.1: Anatomy of the eye.

The purpose of the experiment is to estimate a quadratic model in the CO_2 level that explains the variations in corneal hydration control. Two observations, one for each eye, are made for every human subject. If we denote the number of subjects involved in the study by b, then the total number of observations in the study is equal to $n = 2b$. Of course, the two observations made for one subject are likely to be correlated, so that each subject serves as a block of two correlated observations. The number of subjects available lies between 30 and 60.

5.3   Model

Let us now denote by y a measure of the corneal hydration control and by x the level of the CO_2 treatment applied. The model of interest can then be written as

$$y_{ij} = \beta_0 + \beta_1 x + \beta_2 x^2, \qquad (5.1)$$

where β_0, β_1 and β_2 represent the intercept, the linear effect and the quadratic effect respectively. The statistical model corresponding to the experiment takes into account the random variation in each observation and the fact that each subject in the study is different. Therefore, the statistical model contains a random block effect for each subject in the study and an error term reflecting the random variation in each observation. The response of the jth observation for the ith subject can then be written as

$$y_{ij} = \beta_0 + \beta_1 x_{ij} + \beta_2 x_{ij}^2 + \gamma_i + \varepsilon_{ij}, \qquad (5.2)$$

where x_{ij} is the jth CO_2 level applied to the ith subject, γ_i is the random effect corresponding to the ith subject and ε_{ij} is the random error. Since two measurements are made for each subject, the block size of the experiment is equal to two and the index j can only take the values 1 or 2. In matrix notation, the model becomes

$$\mathbf{y} = \mathbf{X}\boldsymbol{\beta} + \mathbf{Z}\boldsymbol{\gamma} + \boldsymbol{\varepsilon}, \qquad (5.3)$$

where $\mathbf{y}$ is a vector of n observations on the corneal hydration control, the vector $\boldsymbol{\beta}$ contains the three unknown fixed parameters, the vector $\boldsymbol{\gamma} = [\ \gamma_1\ \gamma_2\ \cdots\ \gamma_b\ ]'$ contains the b random block effects and $\boldsymbol{\varepsilon}$ is an n-dimensional random error vector. The matrices $\mathbf{X}$ and $\mathbf{Z}$ are known and have dimension $n \times 3$ and $n \times b$ respectively. The n rows of $\mathbf{X}$ contain a one corresponding to the intercept, the CO_2 level for each observation and its square. The matrix $\mathbf{Z}$ assigns the treatments to the subjects. When the observations are grouped per subject, $\mathbf{Z}$ is of the form

$$\mathbf{Z} = \mathrm{diag}[1_2, 1_2, \ldots, 1_2], \qquad (5.4)$$

where $\mathbf{1}_2$ is a 2-dimensional vector of ones. It is assumed that

$$E(\boldsymbol{\varepsilon}) = \mathbf{0}_n \text{ and } \mathrm{Cov}(\boldsymbol{\varepsilon}) = \sigma_\varepsilon^2 \mathbf{I}_n, \tag{5.5}$$

$$E(\boldsymbol{\gamma}) = \mathbf{0}_b \text{ and } \mathrm{Cov}(\boldsymbol{\gamma}) = \sigma_\gamma^2 \mathbf{I}_b, \tag{5.6}$$

$$\text{and } \mathrm{Cov}(\boldsymbol{\gamma}, \boldsymbol{\varepsilon}) = \mathbf{0}_{b \times n}. \tag{5.7}$$

Under these assumptions, the variance-covariance matrix of the observations $\mathrm{Cov}(\mathbf{y})$ can be written as

$$\mathbf{V} = \mathrm{diag}[\tilde{\mathbf{V}}, \tilde{\mathbf{V}}, \dots, \tilde{\mathbf{V}}], \tag{5.8}$$

where

$$
\begin{aligned}
\tilde{\mathbf{V}} &= \begin{bmatrix} \sigma_\varepsilon^2 + \sigma_\gamma^2 & \sigma_\gamma^2 \\ \sigma_\gamma^2 & \sigma_\varepsilon^2 + \sigma_\gamma^2 \end{bmatrix}, \\
&= \sigma_\varepsilon^2 \begin{bmatrix} 1+\eta & \eta \\ \eta & 1+\eta \end{bmatrix},
\end{aligned} \tag{5.9}
$$

and $\eta = \sigma_\gamma^2/\sigma_\varepsilon^2$ is a measure for the extent to which observations within the same group are correlated. The larger η, the more the observations within one group are correlated. In the optometry experiment, it is expected that σ_γ^2 will be substantially larger than σ_ε^2, or, equivalently, that η will be substantially larger than one.

When the random error terms as well as the block effects are normally distributed, the maximum likelihood estimator of the unknown model parameter $\boldsymbol{\beta}$ in (5.3) is the generalized least squares (GLS) estimator (3.9). The variance-covariance matrix of the estimators is given by (3.10) and the information matrix is given by (3.13).

Using Theorem 18.2.8 of Harville (1997), we have that

$$\tilde{\mathbf{V}}^{-1} = \frac{1}{\sigma_\varepsilon^2}(\mathbf{I}_2 - c\mathbf{1}_2\mathbf{1}_2'), \tag{5.10}$$

where $c = \eta/(1 + 2\eta)$, and since $\mathbf{V}$ is block diagonal,

$$
\begin{aligned}
\mathbf{M} &= \sum_{i=1}^b \mathbf{X}_i' \tilde{\mathbf{V}}^{-1} \mathbf{X}_i \\
&= \frac{1}{\sigma_\varepsilon^2} \sum_{i=1}^b \mathbf{X}_i'(\mathbf{I}_2 - c\mathbf{1}_2\mathbf{1}_2')\mathbf{X}_i, \\
&= \frac{1}{\sigma_\varepsilon^2}\left\{ \sum_{i=1}^b (\mathbf{X}_i'\mathbf{X}_i - c\mathbf{X}_i'\mathbf{1}_2\mathbf{1}_2'\mathbf{X}_i) \right\}, \\
&= \frac{1}{\sigma_\varepsilon^2}\left\{ \mathbf{X}'\mathbf{X} - \sum_{i=1}^b c(\mathbf{X}_i'\mathbf{1}_2)(\mathbf{X}_i'\mathbf{1}_2)' \right\},
\end{aligned} \tag{5.11}
$$

where $\mathbf{X}_i$ is the part of $\mathbf{X}$ corresponding to the ith subject. The problem of designing the optometry experiment consists of choosing the CO_2 levels to be applied to the b subjects. In other words, the matrices $\mathbf{X}$ and $\mathbf{Z}$ have to be determined. As in the previous chapters, the $\mathcal{D}$-optimality criterion will be used to compare alternative design options. The $\mathcal{D}$-optimal design maximizes the determinant of the information matrix (5.11). The problem of finding $\mathcal{D}$-optimal designs for the optometry experiment has already received attention by Chasalow (1992), Cheng (1995) and Atkins and Cheng (1999). Chasalow (1992) used complete enumeration to find the best possible exact designs with the levels -1, 0 and +1 for several numbers of subjects b. Cheng (1995) and Atkins and Cheng (1999) use an approximate theory to derive optimal continuous designs for the optometry experiment. We examine these results in detail in Section 5.4 and in Section 5.5.1. In Section 5.5.2, we derive exact $\mathcal{D}$-optimal designs with b blocks of two observations for the estimation of the quadratic model (5.3). The resulting designs are much more efficient than the three-level designs derived by Chasalow (1992). In the sequel of the chapter, we denote the two treatments given to a subject by $(x_{i1}; x_{i2})$. The CO_2 level x is represented in coded form: its minimal and maximal value will be denoted by -1 and 1 respectively, hence $x_{ij} \in [-1, 1]$ $(i = 1, 2, \ldots, b; j = 1, 2)$.

5.4   Continuous $\mathcal{D}$-optimal designs

Cheng (1995) and Atkins and Cheng (1999) derive continuous $\mathcal{D}$-optimal designs for the optometry experiment. They show that the continuous $\mathcal{D}$-optimal design is supported on the blocks[1] $(1; -\alpha_\eta)$, $(-1; \alpha_\eta)$ and $(-1; 1)$, where $\alpha_\eta \geq 0$, with weights w_η, w_η and $1 - 2w_\eta$ respectively. Both α_η and w_η are increasing functions of η. Cheng (1995) shows that $\alpha_\eta \rightarrow 0$ and $\varepsilon_\eta \rightarrow 2/3$ when η approaches zero. As an illustration, optimal values of α_η and w_η for several values of η are given in Table 5.1.

The continuous optimal designs for the optometry experiment possess four different factor levels: -1, $-\alpha_\eta$, α_η and 1. This is different from the continuous $\mathcal{D}$-optimal design for a model without block effects, which is supported on the levels -1, 0 and 1. It also turns out that the three blocks of the experiment do not receive equal weights when η is strictly positive. The blocks $(1; -\alpha_\eta)$ and $(-1; \alpha_\eta)$ both receive more weight than the block $(-1; 1)$. This is increasingly so when η increases. Finally, note that the pace with which α_η and w_η increase becomes very small for large values of η.

---

[1] While a continuous design for an unblocked experiment is represented by a measure on the set of design points, a continuous design for a blocked experiment is represented by a measure on the set of blocks in the experiment.

Table 5.1: Values of α_η and w_η in the continuous $\mathcal{D}$-optimal design for the optometry experiment.

η	α_η	w_η	$1 - 2w_\eta$
0	0.000	0.333	0.333
0.1	0.029	0.334	0.331
0.25	0.059	0.338	0.324
0.5	0.093	0.345	0.311
0.75	0.115	0.351	0.299
1	0.131	0.356	0.288
2	0.167	0.370	0.260
5	0.202	0.386	0.228
10	0.218	0.394	0.212
100	0.234	0.403	0.193
∞	0.236	0.405	0.191

For the computation of continuous designs, it is assumed that an infinitely large number of subjects is available. In practice, however, this is not the case. In the next section, we will compute exact $\mathcal{D}$-optimal designs for the optometry experiment and compare them to the designs obtained by rounding the $\mathcal{D}$-optimal continuous design.

5.5   Exact $\mathcal{D}$-optimal designs

Chasalow (1992) computes the best possible exact designs with three factor levels, namely -1, 0 and 1, for the optometry experiment. His results are described in the first part of this section. In the second part, we show that the three-level designs can be improved to a large extent by using other factor levels as well.

5.5.1   Three-level designs

Chasalow (1992) uses complete enumeration to find the $\mathcal{D}$-optimal three-level designs for the optometry experiment for several values of b. It turns out that the optimal three-level designs are supported on three different blocks: $(1;0)$, $(-1;0)$ and $(-1;1)$. If b is a multiple of three, then each of the blocks is used $b/3$ times in the $\mathcal{D}$-optimal design. In that case, the $\mathcal{D}$-optimal design is a balanced incomplete block design. If b is not a multiple of three, the three types of blocks are used with frequencies as equal as possible. Cheng (1995) shows that the designs derived by Chasalow are $\mathcal{D}$-optimal among all minimum support designs —that is the set of designs with p distinct design points— for any strictly positive η.

5.5.2 $\mathcal{D}$-optimal designs

The three-level designs described in Section 5.5.1 are not optimal when the number of support points is allowed to be more than the number of fixed model parameters p. In this section, we show that the $\mathcal{D}$-optimal designs for the optometry experiment possess four factor levels. The $\mathcal{D}$-optimal designs are computed by combining the blocking algorithm of Goos and Vandebroek (2001a) and the adjustment algorithm of Donev and Atkinson (1988). Both algorithms are discussed in the previous chapter. The algorithm of Goos and Vandebroek (2001a) produces the $\mathcal{D}$-optimal three-level designs described in Section 5.5.1 when the default set of the candidate points -1, 0 and +1 is used. However, it produces substantially better designs when a set of 21 equally spaced points between -1 and 1 is used. The resulting designs can be further improved by applying the adjustment algorithm. For small numbers of b, we will use analytical results to evaluate this approach.

Designs with two blocks

First, consider the problem of designing an optometry experiment with two blocks of two observations. Hence $b = 2$ and $n = 4$. When $\eta = 0$, the design problem reduces to the computation of a 4-point $\mathcal{D}$-optimal completely randomized design, which has observations in the points -1, 0 and 1, one of which is duplicated. Typically, the symmetric design with the duplicated center point will be preferred because the linear effect can then be estimated independently of the intercept and the quadratic effect. When $\eta > 0$, the $\mathcal{D}$-optimal designs generated by the algorithmic approach have four different factor levels: -1, $-a_\eta$, a_η and 1, where $a_\eta > 0$. The first block of the optimal design contains the points -1 and a_η. The second block contains the points $-a_\eta$ and 1. It turns out that a smaller η results in a smaller a_η. When $\eta \to 0$, $a_\eta \to 0$. A similar result was found for continuous designs. We have displayed the optimal design points for several values of η in Figure 5.2. The figure clearly shows that a_η increases with η. Now, we will show how the exact $\mathcal{D}$-optimal values for a_η can be computed analytically. It will also be shown that a_η approaches $1/3$ when $\eta \to \infty$.

For notational simplicity, assume without loss of generality that $\sigma_\varepsilon^2 = 1$. Substituting $b = 2$ in (5.11), we then have

$$\mathbf{X}'\mathbf{V}^{-1}\mathbf{X} = \mathbf{X}'\mathbf{X} - c\sum_{i=1}^{2}(\mathbf{X}_i'\mathbf{1}_2)(\mathbf{X}_i'\mathbf{1}_2)'. \qquad (5.12)$$

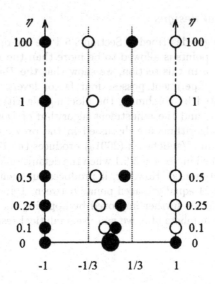

Figure 5.2: $\mathcal{D}$-optimal design points for the optometry experiment when $b = 2$. A • indicates a design point from the first block, a ○ indicates a design point from the second block.

For the design problem at hand, the optimal design is of the form

$$\mathbf{X} = \begin{bmatrix} \mathbf{X}_1 \\ \mathbf{X}_2 \end{bmatrix} = \begin{bmatrix} 1 & -1 & 1 \\ 1 & a_\eta & a_\eta^2 \\ 1 & -a_\eta & a_\eta^2 \\ 1 & 1 & 1 \end{bmatrix},$$

with

$$\mathbf{X}_1 = \begin{bmatrix} 1 & -1 & 1 \\ 1 & a_\eta & a_\eta^2 \end{bmatrix} \text{ and } \mathbf{X}_2 = \begin{bmatrix} 1 & -a_\eta & a_\eta^2 \\ 1 & 1 & 1 \end{bmatrix}.$$

Therefore,

$$\mathbf{X}'\mathbf{X} = \begin{bmatrix} 4 & 0 & 2(1 + a_\eta^2) \\ 0 & 2(1 + a_\eta^2) & 0 \\ 2(1 + a_\eta^2) & 0 & 2(1 + a_\eta^4) \end{bmatrix},$$

$$\mathbf{X}_1'\mathbf{1}_2 = \begin{bmatrix} 2 \\ a_\eta - 1 \\ 1 + a_\eta^2 \end{bmatrix} \text{ and } \mathbf{X}_2'\mathbf{1}_2 = \begin{bmatrix} 2 \\ 1 - a_\eta \\ 1 + a_\eta^2 \end{bmatrix},$$

$$(\mathbf{X}_1'\mathbf{1}_2)(\mathbf{X}_1'\mathbf{1}_2)' = \begin{bmatrix} 4 & 2(a_\eta - 1) & 2(1 + a_\eta^2) \\ 2(a_\eta - 1) & (a_\eta - 1)^2 & (a_\eta - 1)(1 + a_\eta^2) \\ 2(1 + a_\eta^2) & (a_\eta - 1)(1 + a_\eta^2) & (1 + a_\eta^2)^2 \end{bmatrix},$$

$$(\mathbf{X}_2'\mathbf{1}_2)(\mathbf{X}_2'\mathbf{1}_2)' = \begin{bmatrix} 4 & 2(1-a_\eta) & 2(1+a_\eta^2) \\ 2(1-a_\eta) & (a_\eta-1)^2 & (1-a_\eta)(1+a_\eta^2) \\ 2(1+a_\eta^2) & (1-a_\eta)(1+a_\eta^2) & (1+a_\eta^2)^2 \end{bmatrix},$$

and

$$\sum_{i=1}^{2}(\mathbf{X}_i'\mathbf{1}_2)(\mathbf{X}_i'\mathbf{1}_2)' = \begin{bmatrix} 8 & 0 & 4(1+a_\eta^2) \\ 0 & 2(a_\eta-1)^2 & 0 \\ 4(1+a_\eta^2) & 0 & 2(1+a_\eta^2)^2 \end{bmatrix}.$$

Substituting these results in (5.12), yields the following information matrix:

$$\mathbf{X}'\mathbf{V}^{-1}\mathbf{X} = 2\begin{bmatrix} 2-4c & 0 & (1-2c)(1+a_\eta^2) \\ 0 & 1+a_\eta^2-c(a_\eta-1)^2 & 0 \\ (1-2c)(1+a_\eta^2) & 0 & 1+a_\eta^4-c(1+a_\eta^2)^2 \end{bmatrix}.$$

The $\mathcal{D}$-criterion value is then given by

$$|\mathbf{X}'\mathbf{V}^{-1}\mathbf{X}| = 2^3(1-a_\eta^2-a_\eta^4+a_\eta^6-3c+2a_\eta c+3a_\eta^2 c-4a_\eta^3 c+3a_\eta^4 c+2a_\eta^5 c$$
$$- 3a_\eta^6 c+2c^2-4a_\eta c^2-2a_\eta^2 c^2+8a_\eta^3 c^2-2a_\eta^4 c^2-4a_\eta^5 c^2+2a_\eta^6 c^2).$$

This determinant reaches a maximum when the first derivative with respect to a_η

$$\frac{d|\mathbf{X}'\mathbf{V}^{-1}\mathbf{X}|}{da_\eta} = 2^3(-2a_\eta - 4a_\eta^3 + 6a_\eta^5 + 2c + 6a_\eta c - 12a_\eta^2 c + 12a_\eta^3 c + 10a_\eta^4 c$$
$$- 18a_\eta^5 c - 4c^2 - 4a_\eta c^2 + 24a_\eta^2 c^2 - 8a_\eta^3 c^2 - 20a_\eta^4 c^2 + 12a_\eta^5 c^2)$$

equals zero and when the second derivative with respect to a_η

$$\frac{d^2|\mathbf{X}'\mathbf{V}^{-1}\mathbf{X}|}{da_\eta^2} = 2^3(-2 - 12a_\eta^2 + 30a_\eta^4 + 6c - 24a_\eta c + 36a_\eta^2 c + 40a_\eta^3 c$$
$$- 90a_\eta^4 c - 4c^2 + 48a_\eta c^2 - 24a_\eta^2 c^2 - 80a_\eta^3 c^2 + 60a_\eta^4 c^2)$$

is strictly negative. These conditions are fulfilled for

$$a_\eta = \frac{1}{9(1-c)}\left(-5c + \frac{\kappa\sqrt[3]{2}}{\sqrt[3]{\lambda+\sqrt{4\kappa^3+\lambda^2}}} - \frac{\sqrt[3]{\lambda+\sqrt{4\kappa^3+\lambda^2}}}{\sqrt[3]{2}}\right), \quad (5.13)$$

with

$$\kappa = 9 - 18c - 16c^2 \text{ and } \lambda = -243(1-c)^2 c + 135(1-c)(c-1)c + 250c^3.$$

Substituting different values for c in (5.13) yields the corresponding optimal value for a_η. For example, when $\eta = 1$, $c = 1/3$, $\kappa = 11/9$ and $\lambda = -47 + 7/27$. As a result, the optimal value for a_η is 0.266218. We have performed similar computations for other values of η. The results are given in Table 5.2. When $\eta \to \infty$, $c \to 1/2$, $\kappa \to -4$ and $\lambda \to -16$. As a consequence, $a_\eta \to 1/3$ when $\eta \to \infty$.

Table 5.2: $\mathcal{D}$-optimal values for a_η when two blocks of size two are used for quadratic regression on one variable.

η	c	κ	λ	a_η
0.1	0.0833	7.3889	-26.3241	0.085685
0.25	0.1667	5.5556	-42.5926	0.161359
0.5	0.2500	3.5000	-49.2500	0.220333
1	0.3333	1.2222	-46.7407	0.266218
2	0.4000	-0.7600	-38.4320	0.296215
5	0.4545	-2.4876	-27.6409	0.317454
10	0.4762	-3.1996	-22.3928	0.325202
100	0.4975	-3.9155	-16.6980	0.332502
∞	0.5000	-4.0000	-16.0000	0.333333

Designs with three blocks

Now, consider the problem of designing an optometry experiment with three blocks. When $\eta = 0$, the $\mathcal{D}$-optimal design has two observations in the points -1, 0 and 1. When $\eta > 0$, the algorithmic approach again produces designs with four different factor levels: -1, $-b_\eta$, b_η and 1, where $b_\eta > 0$. The first block of the optimal design contains the points -1 and 1. The second block contains the points -1 and b_η and the third block contains the points $-b_\eta$ and 1. It turns out that b_η increases with η. This result does not come as a surprise in view of the results of Cheng (1995), who proves that the $\mathcal{D}$-optimal continuous design for the design problem at hand is supported on three blocks with a similar structure (see Section 5.4). As in the case where only two blocks were available, the exact $\mathcal{D}$-optimal values for b_η can be computed analytically. Substituting $\sigma_\varepsilon^2 = 1$ and $b = 3$ in (5.11), we obtain

$$\mathbf{X}'\mathbf{V}^{-1}\mathbf{X} = \mathbf{X}'\mathbf{X} - c\sum_{i=1}^{3}(\mathbf{X}_i'\mathbf{1}_2)(\mathbf{X}_i'\mathbf{1}_2)'. \qquad (5.14)$$

For the design problem at hand, the optimal design is of the form

$$\mathbf{X} = \begin{bmatrix} \mathbf{X}_1 \\ \mathbf{X}_2 \\ \mathbf{X}_3 \end{bmatrix} = \begin{bmatrix} 1 & -1 & 1 \\ 1 & 1 & 1 \\ 1 & -1 & 1 \\ 1 & b_\eta & b_\eta^2 \\ 1 & -b_\eta & b_\eta^2 \\ 1 & 1 & 1 \end{bmatrix},$$

with

$$\mathbf{X}_1 = \begin{bmatrix} 1 & -1 & 1 \\ 1 & 1 & 1 \end{bmatrix}, \quad \mathbf{X}_2 = \begin{bmatrix} 1 & -1 & 1 \\ 1 & b_\eta & b_\eta^2 \end{bmatrix} \text{ and } \mathbf{X}_3 = \begin{bmatrix} 1 & -b_\eta & b_\eta^2 \\ 1 & 1 & 1 \end{bmatrix}.$$

Therefore,

$$\mathbf{X}'\mathbf{X} = \begin{bmatrix} 6 & 0 & 2(2+b_\eta^2) \\ 0 & 2(2+b_\eta^2) & 0 \\ 2(2+b_\eta^2) & 0 & 2(2+b_\eta^4) \end{bmatrix},$$

$$\mathbf{X}_1'1_2 = \begin{bmatrix} 2 \\ 0 \\ 2 \end{bmatrix}, \quad \mathbf{X}_2'1_2 = \begin{bmatrix} 2 \\ b_\eta - 1 \\ 1 + b_\eta^2 \end{bmatrix} \text{ and } \mathbf{X}_3'1_2 = \begin{bmatrix} 2 \\ 1 - b_\eta \\ 1 + b_\eta^2 \end{bmatrix},$$

$$(\mathbf{X}_1'1_2)(\mathbf{X}_1'1_2)' = \begin{bmatrix} 4 & 0 & 4 \\ 0 & 0 & 0 \\ 4 & 0 & 4 \end{bmatrix},$$

$$(\mathbf{X}_2'1_2)(\mathbf{X}_2'1_2)' = \begin{bmatrix} 4 & 2(b_\eta - 1) & 2(1+b_\eta^2) \\ 2(b_\eta - 1) & (b_\eta - 1)^2 & (b_\eta - 1)(1+b_\eta^2) \\ 2(1+b_\eta^2) & (b_\eta - 1)(1+b_\eta^2) & (1+b_\eta^2)^2 \end{bmatrix},$$

$$(\mathbf{X}_3'1_2)(\mathbf{X}_3'1_2)' = \begin{bmatrix} 4 & 2(1-b_\eta) & 2(1+b_\eta^2) \\ 2(1-b_\eta) & (b_\eta - 1)^2 & (1-b_\eta)(1+b_\eta^2) \\ 2(1+b_\eta^2) & (1-b_\eta)(1+b_\eta^2) & (1+b_\eta^2)^2 \end{bmatrix},$$

and

$$\sum_{i=1}^{3}(\mathbf{X}_i'1_2)(\mathbf{X}_i'1_2)' = \begin{bmatrix} 12 & 0 & 4(2+b_\eta^2) \\ 0 & 2(b_\eta - 1)^2 & 0 \\ 4(2+b_\eta^2) & 0 & 4+2(1+b_\eta^2)^2 \end{bmatrix}.$$

Substituting these results in (5.14), yields the following information matrix:

$$\mathbf{X}'\mathbf{V}^{-1}\mathbf{X} = 2\begin{bmatrix} 3-6c & 0 & (1-2c)(2+b_\eta^2) \\ 0 & 2+b_\eta^2 - c(b_\eta - 1)^2 & 0 \\ (1-2c)(2+b_\eta^2) & 0 & 2+b_\eta^4 - c\{2+(1+b_\eta^2)^2\} \end{bmatrix}.$$

The determinant of this matrix is maximized in

$$b_\eta = \frac{1}{9(1-c)}\left(-5c + \frac{\theta\sqrt[3]{2}}{\sqrt[3]{\tau + \sqrt{4\theta^3 + \tau^2}}} - \frac{\sqrt[3]{\tau + \sqrt{4\theta^3 + \tau^2}}}{\sqrt[3]{2}}\right), \quad (5.15)$$

with

$$\theta = 27 - 36c - 16c^2 \text{ and } \tau = -243(1-c)^2c + 135(1-c)(c-3)c + 250c^3.$$

In Table 5.3, $\mathcal{D}$-optimal values for b_η are given for several values of η. The values found are different from those found by Cheng (1995) for the $\mathcal{D}$-optimal continuous design. This is because the $\mathcal{D}$-optimal continuous design does not have an equal weight on the three blocks whereas in the discrete case, the weight of each block is equal to one.

Table 5.3: $\mathcal{D}$-optimal values for b_η when three blocks of size two are used for the optometry experiment.

η	c	θ	τ	b_η
0.1	0.0833	23.8889	-46.9491	0.028434
0.25	0.1667	20.5556	-80.0926	0.057676
0.5	0.2500	17.000	-99.8750	0.086936
1	0.3333	13.2222	-106.7407	0.115506
2	0.4000	10.0400	-103.2320	0.137503
5	0.4545	7.3306	-94.5830	0.154793
10	0.4762	6.2290	-89.7398	0.161464
100	0.4975	5.1293	-84.1963	0.167924
∞	0.5000	5.0000	-83.5000	0.168663

Table 5.4: Comparison of the $\mathcal{D}$-optimal values for a_η and b_η and the values computed by the adjustment algorithm (A.A.).

η	a_η		b_η	
	Exact	A.A.	Exact	A.A.
0.1	0.085685	0.085000	0.028434	0.027500
0.25	0.161359	0.160000	0.057676	0.057500
0.5	0.220333	0.220000	0.086936	0.087500
1	0.266218	0.267500	0.115506	0.115000
2	0.296215	0.297500	0.137503	0.137500
5	0.317454	0.317500	0.154793	0.155000
10	0.325202	0.325000	0.161464	0.162500
100	0.332502	0.332500	0.167924	0.167500

The algorithmic approach produces values for a_η and b_η that closely approximate the ones analytically derived and displayed in Tables 5.2 and 5.3. This is illustrated in Table 5.4.

Designs with four or five blocks

The structure of the $\mathcal{D}$-optimal designs with two or three blocks of size two for quadratic regression on one variable appears to be constant for all values of η. As is demonstrated by the optimal designs displayed in Table 5.5, this is not always the case when b is larger. In the table, the numbers r_i represent the number of times the ith type of block is used in the experiment. Consider for example the case where the number of blocks b is equal to four. When η is small, two equivalent $\mathcal{D}$-optimal designs are supported on three different blocks. One design is supported on the blocks $(-1; c_\eta)$, $(-d_\eta; 1)$ and $(-1; 1)$, with $0 < c_\eta < d_\eta$. The block $(-1; c_\eta)$ appears twice in the optimal design, while the other two appear only once. The

Table 5.5: $\mathcal{D}$-optimal designs with four or five blocks.

b	η	r_1	Block 1	r_2	Block 2	r_3	Block 3
4	0.1	2	(-1;0.025)	1	(-0.050;1)	1	(-1;1)
		1	(-1;0.050)	2	(-0.025;1)	1	(-1;1)
	0.5	2	(-1;0.080)	1	(-0.145;1)	1	(-1;1)
		1	(-1;0.145)	2	(-0.080;1)	1	(-1;1)
	1	2	(-1;0.106)	1	(-0.185;1)	1	(-1;1)
		1	(-1;0.185)	2	(-0.106;1)	1	(-1;1)
	5	2	(-1;0.318)	2	(-0.318;1)		
	10	2	(-1;0.325)	2	(-0.325;1)		
5	0.1	2	(-1;0.015)	1	(-0.030;1)	2	(-1;1)
		1	(-1;0.030)	2	(-0.015;1)	2	(-1;1)
	0.5	2	(-1;0.129)	2	(-0.129;1)	1	(-1;1)
	1	2	(-1;0.168)	2	(-0.168;1)	1	(-1;1)
	5	2	(-1;0.215)	2	(-0.215;1)	1	(-1;1)
	10	2	(-1;0.223)	2	(-0.223;1)	1	(-1;1)

mirror image of this design, obtained by multiplying its factor levels by -1, is equivalent. It turns out that both c_η and d_η are increasing functions of η. When η is large, the $\mathcal{D}$-optimal designs with four blocks are supported on two different blocks $(-1; f_\eta)$ and $(-f_\eta; 1)$, with $0 < f_\eta$ and f_η an increasing function of η. Both blocks are replicated twice.

When $b = 5$ and η is small, the $\mathcal{D}$-optimal designs are supported on the blocks $\pm(-1; g_\eta)$, $\pm(-h_\eta; 1)$ and $\pm(-1; 1)$, with $0 < g_\eta < h_\eta$ and both g_η and h_η increasing functions of η. While the block $\pm(-h_\eta; 1)$ is used only once, the blocks $\pm(-1; g_\eta)$ and $\pm(-1; 1)$ are used twice. When $b = 5$ and η is not too small, the $\mathcal{D}$-optimal design has two blocks of type $(-1; q_\eta)$, two blocks of type $(-q_\eta; 1)$ and one block $(-1; 1)$, where $q_\eta > 0$ and increases with η. Some $\mathcal{D}$-optimal values of g_η, h_η and q_η are given in Table 5.5.

Designs with six blocks

When six blocks are available for the optometry experiment, the $\mathcal{D}$-optimal design is given by two replicates of the optimal design for $b = 3$. However, this statement cannot be generalized to other multiples of three.

Large numbers of blocks

The structure of the $\mathcal{D}$-optimal designs with two or three blocks of size two for quadratic regression on one variable appears to be constant for all values of η. As was already illustrated for experiments with four or five blocks, this is no longer the case when b is larger. Another observation is that the optimal designs begin to resemble the continuous $\mathcal{D}$-optimal designs when the number of blocks b further increases. The $\mathcal{D}$-optimal designs are then supported on blocks of type $(-1, s_\eta)$ and $(-t_\eta, 1)$, with $0 < s_\eta$ and $0 < t_\eta$, and on the block $(-1, 1)$. Not surprisingly, s_η and t_η are increasing functions of η. In the optimal designs, the first two blocks are used with frequencies as equal as possible. Therefore, the absolute difference between r_1 and r_2 is at most one. In all cases where r_1 is equal to r_2, s_η and t_η are equal as well. In cases where r_1 and r_2 are different, $s_\eta < t_\eta$ when $r_1 = r_2 + 1$ and $s_\eta > t_\eta$ when $r_1 = r_2 - 1$. Some examples of exact $\mathcal{D}$-optimal designs for large values of b are given in the left panel of Table 5.6.

Efficiency comparisons for large numbers of subjects

Comparing the exact $\mathcal{D}$-optimal designs in Table 5.6 with the three-level designs in the middle panel of the table in terms of $\mathcal{D}$-efficiency shows that the former are more efficient than the latter, especially for large degrees of correlation. For $\eta = 0.1$, the three-level designs are 0.04% less efficient than the $\mathcal{D}$-optimal ones. However, they are 2.25% less efficient when $\eta = 10$. This is not unexpected because the $\mathcal{D}$-optimal designs for small η strongly resemble the three-level designs, while both the design points and the numbers of replicates of the blocks are completely different for larger values of η. The relative performance of the designs is independent of the number of subjects available.

Rounding the continuous $\mathcal{D}$-optimal design, that is setting $r_1 = r_2 = [bw_\eta]$ and $r_3 = b - r_1 - r_2$, does not yield the exact $\mathcal{D}$-optimal design. Firstly, the factor levels obtained from the continuous design are different from those of the exact $\mathcal{D}$-optimal design. Secondly, rounding the weights w_η and $1 - 2w_\eta$ of the continuous design does not always produce the optimal numbers of replicates r_i. Suppose we would like to construct a design with 36 blocks from the $\mathcal{D}$-optimal continuous design for $\eta = 0.5$. As can be seen in Table 5.1, the weight w_η assigned to the blocks of type $(-1; 0.093)$ and $(-0.093; 1)$ is 0.345. In a design with 36 blocks, this type of block should thus be used $36 \times 0.345 = 12.42$ times. Rounding this value to the nearest integer gives us $r_1 = r_2 = 12$, and hence $r_3 = 12$. The resulting $\mathcal{D}$-criterion value is nearly identical to that of the $\mathcal{D}$-optimal designs with $r_1 = 13$, $r_2 = 12$ and $r_3 = 11$ given in Table 5.6. As a result, rounding the continuous $\mathcal{D}$-optimal design produces a design that is only slightly less efficient than the $\mathcal{D}$-optimal design. This is also the case for other values of b and η, even though the factor levels α_η of the continuous design are different from the optimal levels given in Table 5.6.

Table 5.6: $\mathcal{D}$-optimal designs with a large number of blocks for the optometry experiment. All designs are supported on three types of blocks. Two of the block types are different for the three design options, while the third block type is $(-1;1)$ for every type. For the three-level and the continuous designs, the second block is obtained from the first by multiplying its levels by -1.

DESIGN OPTIONS

DESIGN PROBLEM		EXACT D-OPTIMAL DESIGNS					THREE-LEVEL DESIGNS				CONTINUOUS D-OPTIMAL DESIGNS			
b	η	r_1	r_2	r_3	Block 1	Block 2	$r_1=r_2$	r_3	Block 1	rel.eff.	$r_1=r_2$	r_3	Block 1	rel.eff.
36	0.1	12	12	12	(-1;0.028)	(-0.028;1)	12	12	(-1;0)	0.999587	12	12	(-1;0.029)	0.999999
	0.5	13	12	11	(-1;0.091)	(-0.098;1)	12	12	(-1;0)	0.995755	12	12	(-1;0.093)	0.999898
	1	13	13	10	(-1;0.135)	(-0.135;1)	12	12	(-1;0)	0.991312	13	10	(-1;0.131)	0.999990
	5	14	14	8	(-1;0.205)	(-0.205;1)	12	12	(-1;0)	0.980342	14	8	(-1;0.202)	0.999995
	10	14	14	8	(-1;0.212)	(-0.212;1)	12	12	(-1;0)	0.977565	14	8	(-1;0.218)	0.999975
48	0.1	16	16	16	(-1;0.028)	(-0.028;1)	16	16	(-1;0)	0.999588	16	16	(-1;0.029)	1.000000
	0.5	17	16	15	(-1;0.090)	(-0.095;1)	16	16	(-1;0)	0.995669	17	14	(-1;0.093)	0.999887
	1	17	17	14	(-1;0.130)	(-0.130;1)	16	16	(-1;0)	0.991267	17	14	(-1;0.131)	0.999999
	5	19	18	11	(-1;0.198)	(-0.205;1)	16	16	(-1;0)	0.980452	19	10	(-1;0.202)	0.999927
	10	19	19	10	(-1;0.219)	(-0.219;1)	16	16	(-1;0)	0.977539	19	10	(-1;0.218)	0.999999
49	0.1	17	16	16	(-1;0.028)	(-0.030;1)	16	17	(-1;0)	0.999572	16	17	(-1;0.029)	0.999954
	0.5	17	17	15	(-1;0.094)	(-0.094;1)	16	17	(-1;0)	0.995479	17	15	(-1;0.093)	0.999999
	1	18	17	14	(-1;0.129)	(-0.135;1)	16	17	(-1;0)	0.991245	17	15	(-1;0.131)	0.999925
	5	19	19	11	(-1;0.204)	(-0.204;1)	16	17	(-1;0)	0.980207	19	11	(-1;0.202)	0.999998
	10	19	19	11	(-1;0.211)	(-0.211;1)	16	17	(-1;0)	0.977459	19	11	(-1;0.218)	0.999965
60	0.1	20	20	20	(-1;0.028)	(-0.028;1)	20	20	(-1;0)	0.999588	20	20	(-1;0.029)	1.000000
	0.5	21	21	18	(-1;0.096)	(-0.096;1)	20	20	(-1;0)	0.995639	21	18	(-1;0.093)	0.999995
	1	21	21	18	(-1;0.127)	(-0.127;1)	20	20	(-1;0)	0.991327	21	18	(-1;0.131)	0.999990
	5	23	23	14	(-1;0.199)	(-0.199;1)	20	20	(-1;0)	0.980344	23	14	(-1;0.202)	0.999995
	10	24	24	12	(-1;0.223)	(-0.223;1)	20	20	(-1;0)	0.977578	24	12	(-1;0.218)	0.999979

5.6   Discussion

A common feature of all exact $\mathcal{D}$-optimal designs for the problem under
consideration is that they possess four different design points. As was illustrated in Figure 5.2, the design points move away from the center point
when the degree of correlation η grows larger. In addition, the number of
times r_3 the block $(-1; 1)$ appears in the exact optimal designs decreases
with η, while the opposite is true for the other blocks. A similar behavior
was encountered when examining the continuous $\mathcal{D}$-optimal designs.

The exact $\mathcal{D}$-optimal designs are substantially more efficient than the best
possible three-level designs, especially for the large degrees of correlation
experienced in practice. It is thus worthwhile to consider other factor levels
than -1, 0 and 1 when designing the optometry experiment. It also turns out
that, although it does not produce the exact $\mathcal{D}$-optimal design, rounding the
continuous $\mathcal{D}$-optimal designs is an excellent design option for this design
problem, so that an algorithmic approach does not add much value in this
example. From a practical point of view, it is also important to stress that
the efficiency of the design obtained in this way does not heavily depend
on η. This is because both the factor levels and the block weights of the
continuous designs do not vary much when η is large as is mostly the case
in practical applications.

6

Constrained Split-Plot Designs

It often happens that all the experimental runs within one group have the same level for one or more factors under investigation. Typically, these factors are hard to change or to control. The resulting design is then called a split-plot design. The groups of a split-plot design are referred to as whole plots and they are divided in so-called sub-plots. The hard-to-change factors are usually called whole plot factors, while the remaining factors are called sub-plot factors.

Experimenters often think of a split-plot experiment as running every sub-plot combination in every whole plot. However, Kempthorne (1952) already describes split-plot designs that do not satisfy this restriction. Similarly, Huang, Chen and Voelkel (1998) and Bingham and Sitter (1999) use the term split-plot design when not all sub-plot combinations occur in each whole plot. In contrast, Letsinger, Myers and Lentner (1996) refer to these type of designs as bi-randomization designs. In this chapter, we will use the term split-plot design (SPD) whether every sub-plot combination is run in every whole plot or not.

The concept of split-plotting, bi-randomization or two-stage randomization is heavily used in industrial experimentation. Cornell (1988) and Kowalski, Cornell and Vining (2002) point out that mixture experiments containing process variables are often of the split-plot type. Letsinger et al. (1996) provide an example of a split-plot experiment from the chemical industry. Bisgaard and Steinberg (1997) demonstrate how split-plotting is applied in prototype experiments. Trinca and Gilmour (2001) compute a split-plot design for a protein extraction experiment. Kowalski and Vining (2001) give an overview of the current literature on the use of split-plot experiments in industry.

6.1   Introduction

In the previous chapters, we have shown that heterogeneous experimental material often forces the experimenter in the direction of a bi-randomization or two-stratum experiment. Another reason to use this form of restricted randomization is that some of the experimental factors are, in some sense, hard to change. For example, heating or cooling down a furnace is time-consuming. In order to save time and money, the experimenters are urged to conduct all the experimental runs with the same furnace temperature successively. In prototype experiments, complete randomization would require building a prototype for each test. Because of cost considerations, this is seldom done. Similarly, in the protein extraction experiment introduced in Example 2 of Section 3.1, a completely randomized design is impractical to carry out. In this experiment, five factors were expected to influence the protein extraction: the feed position for the inflow of a mixture, the feed flow rate, the gas flow rate, the concentration of a protein A and a protein B. Twenty-one days of experimentation were available and three levels were used for each factor. The problem with this experiment was that setting the feed position could be done only once per day because it involved taking apart and reassembling the equipment. By keeping the position fixed during one day, two experimental runs could be performed per day, allowing 42 experimental runs in total.

The protein experiment has two strata. In the first stratum, the experimental runs are assigned to the days. In the second stratum, the two observations performed on one day are arranged in a random order. One factor, the feed position for the inflow of the mixture, is applied to the first stratum. The remaining factors are all applied to the second stratum. As a result, the protein experiment is an example of a split-plot experiment with one whole plot factor and four sub-plot factors. The feed position is the whole plot factor of the experiment, while the other factors are the sub-plot factors. The days are referred to as the whole plots of the experiment and each whole plot is partitioned in two sub-plots. In this experiment, it is clear that the number of whole plots as well as the number of sub-plots within each whole plot are dictated by practical considerations and cannot be chosen at liberty. For this reason, we refer to this type of experiment as a constrained split-plot design.

In the next sections, we describe the response surface model corresponding to a split-plot experiment and discuss the design and the analysis of a split-plot experiment. Next, we derive a number of interesting theoretical results and describe an algorithm to compute $\mathcal{D}$-optimal split-plot designs with given numbers of whole plots and sub-plots. Similarities and differences with the problem of designing blocked experiments will receive attention. Finally, after discussing the computational results, we will develop the best possible design for the protein experiment.

6.2   Model

The set of m experimental variables in a split-plot experiment is partitioned in two subsets. The m_w whole plot variables will be denoted by $w_1, w_2, \ldots, w_{m_w}$ or simply by $\mathbf{w}$. The remaining $m_s = m - m_w$ variables are the sub-plot variables $s_1, s_2, \ldots, s_{m_s}$ or $\mathbf{s}$. In the protein experiment, there is one whole plot factor, namely the feed position, and there are four sub-plot factors: the feed flow rate, the gas flow rate, the concentration of protein A and the concentration of protein B.

First, each of the whole plot factor level combinations of $\mathbf{w}_i$ is assigned randomly to a whole plot, thereby generating the whole plot error variance. The second randomization consists of assigning the combinations of $\mathbf{s}_{ij}$ to the sub-plots, generating the sub-plot error variance. For a polynomial model, the jth observation within the ith whole plot of a split-plot experiment can be written as

$$y_{ij} = \mathbf{f}'(\mathbf{w}_i, \mathbf{s}_{ij})\boldsymbol{\beta} + \gamma_i + \varepsilon_{ij}, \tag{6.1}$$

where $\mathbf{f}(\mathbf{w}_i, \mathbf{s}_{ij})$ represents the polynomial expansion of the whole plot variables and the sub-plot variables, the $p \times 1$ vector $\boldsymbol{\beta}$ contains the p model parameters, γ_i is the random effect of the ith whole plot or the ith whole plot error, and ε_{ij} is the sub-plot error. The difference between the split-plot model and the random block effects model (4.1) lies in the fact that all runs within one whole plot possess the same level for a subset of the experimental factors, namely the whole plot factors. Therefore, the subscript i of the whole plot factors $\mathbf{w}$ refers to the ith whole plot. A full second order split-plot model can be written as

$$\mathbf{f}'(\mathbf{w}_i, \mathbf{s}_{ij})\boldsymbol{\beta} = \beta_0 + \sum_{i=1}^{m_w}(\beta_i^w w_i + \beta_i^{ww} w_i^2) + \sum_{i=1}^{m_s}(\beta_i^s s_i + \beta_i^{ss} s_i^2)$$

$$+ \sum_{i=1}^{m_w}\sum_{j=i+1}^{m_w} \beta_{ij}^{ww} w_i w_j + \sum_{i=1}^{m_s}\sum_{j=i+1}^{m_s} \beta_{ij}^{ss} s_i s_j + \sum_{i=1}^{m_w}\sum_{j=1}^{m_s} \beta_{ij}^{ws} w_i s_j,$$

where β_0 corresponds to the intercept, β_i^w and β_i^{ww} represent the linear and the quadratic effect of the ith whole plot factor, β_i^s and β_i^{ss} are the linear and the quadratic effect of the ith sub-plot factor, β_{ij}^{ww} denotes the interaction effect of the ith and the jth whole plot factor, β_{ij}^{ss} corresponds to the interaction effect of the ith and the jth sub-plot factor, and β_{ij}^{ws} represents the interaction effect of the ith whole plot factor and the jth sub-plot factor. In the sequel, we will refer to β_i^w, β_{ij}^{ww} and β_i^{ww} $(i, j = 1, 2, \ldots, m_w; i \neq j)$ as the whole plot coefficients. Analogously, we will refer to β_i^s, β_{ij}^{ss} and β_i^{ss} $(i, j = 1, 2, \ldots, m_s; i \neq j)$ as the sub-plot coefficients. Finally, we will refer to β_{ij}^{ws} $(i = 1, 2, \ldots, m_w; j = 1, 2, \ldots, m_s)$ as the whole plot by sub-plot interaction coefficients. We will denote the number of whole plot coefficients by p_w and the number of sub-plot coefficients by p_s.

In matrix notation, the model corresponding to a split-plot design is written as

$$\mathbf{Y} = \mathbf{X}\beta + \mathbf{Z}\gamma + \varepsilon, \qquad (6.2)$$

where $\mathbf{X}$ represents the design matrix containing the settings of both the whole plot variables $\mathbf{w}$ and the sub-plot variables $\mathbf{s}$. The matrix $\mathbf{Z}$ is an $n \times b$ matrix of zeroes and ones assigning the n observations to the b whole plots: the (i, j)th entry of $\mathbf{Z}$ is equal to one if the jth observation belongs to the ith whole plot, and zero otherwise. The random effects of the b whole plots are contained within the b-dimensional vector γ, and the random errors are contained within the n-dimensional vector ε. As in the random block effects model, it is assumed that

$$\mathrm{E}(\varepsilon) = \mathbf{0}_n \text{ and } \mathrm{Cov}(\varepsilon) = \sigma_\varepsilon^2 \mathbf{I}_n, \qquad (6.3)$$

$$\mathrm{E}(\gamma) = \mathbf{0}_b \text{ and } \mathrm{Cov}(\gamma) = \sigma_\gamma^2 \mathbf{I}_b, \qquad (6.4)$$

$$\text{and } \mathrm{Cov}(\gamma, \varepsilon) = \mathbf{0}_{b \times n}. \qquad (6.5)$$

The ratio of the whole plot error σ_γ^2 to the sub-plot error σ_ε^2 is again denoted by η, and serves as a measure for the extent to which observations within one whole plot are correlated. We will refer to η as the degree of correlation. Finally, we will denote the size of the ith whole plot by k_i.

6.3   Analysis of a split-plot experiment

As was pointed out in Section 3.3, the unknown fixed parameters β can be estimated by using generalized least squares, while the whole plot error variance σ_γ^2 and the sub-plot error variance σ_ε^2 can be estimated by restricted maximum likelihood (REML). The risks of improperly analyzing a split-plot experiment are pointed out by Box and Jones (1992), Davison (1995) and Ganju and Lucas (1997), who extend the results of Kempthorne (1952). By using a split-plot design, a loss of precision in the estimation of whole

Table 6.1: 3^{4-1} fractional factorial design with one whole plot variable w and three sub-plot variables s_1, s_2 and s_3.

SPD1	SPD2	w	s_1	s_2	s_3
1	1	-1	-1	-1	-1
1	1	-1	0	1	-1
1	1	-1	1	1	1
1	2	-1	-1	0	1
1	2	-1	0	-1	1
1	2	-1	1	0	-1
1	3	-1	-1	1	0
1	3	-1	0	0	0
1	3	-1	1	-1	0
2	4	0	-1	0	-1
2	4	0	0	0	1
2	4	0	1	0	0
2	5	0	-1	1	1
2	5	0	0	-1	-1
2	5	0	1	-1	1
2	6	0	-1	-1	0
2	6	0	0	1	0
2	6	0	1	1	-1
3	7	1	-1	-1	1
3	7	1	0	0	-1
3	7	1	1	-1	-1
3	8	1	-1	1	-1
3	8	1	0	-1	0
3	8	1	1	0	1
3	9	1	-1	0	0
3	9	1	0	1	1
3	9	1	1	1	0

plot coefficients is incurred, while the opposite is true for the sub-plot co-efficients and the whole plot by sub-plot interaction coefficients. The loss of precision in the whole plot coefficients is not illogical since the whole plot factor effects are confounded with the whole plot effects. Analysis of a split-plot experiment as a completely randomized one can therefore lead to erroneously considering whole plot effects as significant and sub-plot effects as insignificant. This is illustrated in Nelson (1985).

In order to demonstrate the loss of precision in the estimation of the whole plot coefficients and the gain in precision for the sub-plot and interaction co-efficients, we have computed the variance-covariance matrix $(\mathbf{X}\mathbf{V}^{-1}\mathbf{X})^{-1}$ of a 3^{4-1} fractional factorial experiment in three cases: the case in which it

Table 6.2: Comparison of the variances of the parameter estimates from a split-plot experiment and a completely randomized design.

	CRD	SPD1	SPD2
Int.	0.3333	0.6667	0.3878
w	0.0556	0.2778	0.1111
s_1	0.0566	0.0278	0.0278
s_2	0.0566	0.0278	0.0390
s_3	0.0566	0.0278	0.0316
w^2	0.1667	0.8333	0.3333
s_1^2	0.1667	0.0833	0.0833
s_2^2	0.1667	0.0833	0.1318
s_3^2	0.1667	0.0833	0.1589
ws_1	0.0889	0.0444	0.0450
ws_2	0.0889	0.0444	0.0649
ws_3	0.0889	0.0444	0.0522
s_1s_2	0.0889	0.0444	0.0749
s_1s_3	0.0889	0.0444	0.0829
s_2s_3	0.0889	0.0444	0.0530

is analyzed as a completely randomized experiment, the case in which it is analyzed as a split-plot experiment with three whole plots, and the case in which it is analyzed as a split-plot experiment with nine whole plots. Let us denote these cases by CRD, SPD1 and SPD2 respectively. For each of them, it was assumed that the total variance was equal to one. For the split-plot experiments, it was assumed that $\sigma_\gamma^2 = \sigma_\varepsilon^2 = 0.5$ and that there is one whole plot variable. The experimental designs of the split-plot experiments are given in Table 6.1. For each case, we have displayed the variances of the parameter estimates in Table 6.2.

Comparing the results of the split-plot experiments SPD1 with those of the completely randomized designs shows that the variances of the estimated pure whole plot coefficients increase substantially. For example, the variance of the estimate for the linear effect of the whole plot factor w increases from 0.0556 to 0.2778. In contrast, the variances of the estimated sub-plot and interaction coefficients are halved. Examining the variances produced by SPD2 allows us to evaluate the impact of an increased number of whole plots. It turns out that the variances of the estimates of the whole plot coefficients are much smaller than those obtained from SPD1. As a result, the negative effect of using a split-plot experiment on the precision of the estimation of the whole plot coefficients can be reduced to a large extent by increasing the number of whole plots. Of course, this result pleads for split-plot experiments with a sufficient number of whole plots.

6.4   Design of a split-plot experiment

Recently, the design of split-plot experiments has received considerable attention in the literature. Huang et al. (1998) and Bingham and Sitter (1999) derive minimum aberration two-level fractional factorial split-plot designs, which are very useful for screening experiments. However, their approach is not useful when the number of runs available is not a power of two, nor when the whole plot size is not a power of two. As a result, many experimental situations exist in which it cannot provide the experimenter with a feasible design. The approach of Huang et al. (1998) and Bingham and Sitter (1999) can be generalized for prime-level designs, which is useful when the experimental factors are qualitative. For lack of alternatives, standard response surface designs are often used to conduct split-plot experiments for fitting second order polynomials. The efficiency of several standard response surface designs under a split-plot error structure was compared by Letsinger et al. (1996). In terms of $\mathcal{D}$-efficiency, the central composite and the Box-Behnken designs turn out to be the best standard designs under a split-plot error structure. However, these standard designs were developed to be used in a completely randomized experiment and not in a split-plot experiment. Therefore, it is probable that these designs are not optimal. In addition, standard response surface designs are not flexible because they are restricted to a few sample sizes and because they cannot be used in experiments with a prespecified number of whole plots of a given size. The design of tailor-made multi-stratum response surface experiments, a special case of which are two-stratum designs, has been treated by Trinca and Gilmour (2001). They develop an algorithm to assign the experimental runs of a given design, often a slightly modified standard response surface design, to the whole plots. The design criteria used, the weighted mean efficiency factor and a weighted $\mathcal{A}$-efficiency criterion, ensure near-orthogonality between the strata. The main drawback of this approach is that the choice of the design points does not take into account the split-plot error structure of the experiment.

In this chapter, we provide the reader with a tool for the efficient design of split-plot experiments with any number of observations n, any number of whole plots b and any whole plot sizes k_i specified by the experimenter. It should be pointed out that the algorithm proposed is able to handle heterogeneous whole plot sizes. In the next section, we derive some interesting theoretical results. In Section 6.6, we describe the design construction algorithm we have developed. As in the previous chapters, the design criterion used is the $\mathcal{D}$-optimality criterion. The aim of the algorithm is thus to find the design matrix $\mathbf{X}$ that maximizes

$$|\mathbf{M}| = |\mathbf{X}'\mathbf{V}^{-1}\mathbf{X}|.$$

It is clear that, in general, the optimal design will depend on the degree of correlation η through the variance-covariance matrix $\mathbf{V}$.

6.5   Some theoretical results

In this section, it will be shown that arranging the runs of a split-plot experiment so that it is crossed, is an optimal approach for a given set of treatments or design points. In addition, the $\mathcal{D}$-optimality of 2^m, 2^{m-f} and Plackett-Burman split-plot designs will be proven.

6.5.1   Optimality of crossed split-plot designs

In a crossed split-plot design, all whole plots have an equal number of sub-plots $k = k_1 = k_2 = \cdots = k_b$, as well as equal levels of the sub-plot variables. As a consequence, the columns of $\mathbf{X}_i$ corresponding to the pure sub-plot terms of model (6.1) are identical across all whole plots. Designs that can be conducted as crossed split-plot designs are full factorial designs, e.g. 2^m designs, and 2^{m-f} designs in which no sub-plot factors are confounded with interactions involving whole plot factors.

Suppose, for example, that 18 runs are available to conduct a split-plot experiment in one whole plot factor and one sub-plot factor, and that the runs have to be arranged in 6 whole plots of size 3. One option is to use a duplicated 3^2 factorial design and to assign its runs to the 6 whole plots. It is clear that the given design possesses three different whole plot levels: -1, 0 and 1. Per whole plot factor level, it contains 6 experimental runs. Since the whole plot size is equal to three, the split-plot experiment will contain two whole plots at each of the three whole plot factor levels. The best possible assignment of the 18 observations in this example results in the crossed split-plot design displayed in Figure 6.1. This is a direct consequence of the result that assigning the treatments of a given design $\mathbf{X}$ to the whole plots so that the resulting split-plot experiment is crossed is an optimal assignment strategy with respect to any generalized optimality criterion. A formal proof of this result is given in Appendix A. As the $\mathcal{D}$-optimality criterion is contained within the class of generalized optimality criteria, arranging the runs of a given design $\mathbf{X}$ so that the resulting split-plot experiment is crossed is the best possible assignment in terms of this criterion. This theoretical result is comforting for practitioners in industry and agriculture who prefer crossed split-plot experiments.

Letsinger et al. (1996) showed that, for a crossed split-plot experiment and a hierarchical model containing an intercept,

$$\mathbf{X}'\mathbf{V}^{-1}\mathbf{X} = \mathbf{G}\mathbf{X}'\mathbf{X}, \tag{6.6}$$

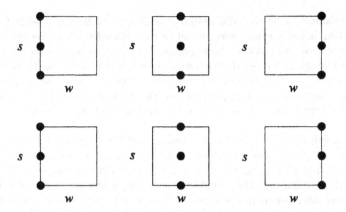

Figure 6.1: Crossed split-plot design with six whole plots of size three for the estimation of a full second order model in one whole plot variable w and one sub-plot variable s.

where

$$
\mathbf{G} = \begin{bmatrix}
(1+k\eta)^{-1} & \mathbf{0}'_{p_w} & \mathbf{0}'_{p_s} & \mathbf{0}'_{p-p_w-p_s} \\
\mathbf{0}_{p_w} & (1+k\eta)^{-1}\mathbf{I}_{p_w} & \mathbf{0}_{p_w \times p_s} & \mathbf{0}_{p_w \times (p-p_w-p_s-1)} \\
-c\mathbf{s} & \mathbf{0}_{p_s \times p_w} & \mathbf{I}_{p_s} & \mathbf{0}_{p_s \times (p-p_w-p_s-1)} \\
\mathbf{0}_{p-p_w-p_s-1} & \mathbf{K} & \mathbf{0}_{(p-p_w-p_s-1) \times p_s} & \mathbf{I}_{p-p_w-p_s-1}
\end{bmatrix},
$$

is a lower triangular matrix of dimension p. In this expression, p_w and p_s represent the number of pure whole plot coefficients and the number of pure sub-plot coefficients, $c = \eta/(1+k\eta)$, and $\mathbf{s}$ and $\mathbf{K}$ are two matrices that depend on the levels of the sub-plot factors used in the experiment. From (6.6), it is clear that the $\mathcal{D}$-criterion value can be written as

$$
|\mathbf{G}||\mathbf{X}'\mathbf{X}|. \tag{6.7}
$$

Although the matrix $\mathbf{G}$ depends on the design matrix $\mathbf{X}$, its determinant does not and is equal to

$$
|\mathbf{G}| = \left(\frac{\sigma_\varepsilon^2}{\sigma_\varepsilon^2 + k\sigma_\gamma^2}\right)^{p_w+1} = \left(\frac{1}{1+k\eta}\right)^{p_w+1}.
$$

From this result, we have that arranging the points of a $\mathcal{D}$-optimal design for the uncorrelated model

$$
\mathbf{Y} = \mathbf{X}\boldsymbol{\beta} + \boldsymbol{\varepsilon}, \tag{6.8}
$$

where $\mathrm{E}(\boldsymbol{\varepsilon}) = \mathbf{0}_n$ and $\mathrm{Cov}(\boldsymbol{\varepsilon}) = \sigma_\varepsilon^2 \mathbf{I}_n$, in a crossed fashion, yields a $\mathcal{D}$-optimal split-plot design within the class of crossed designs with b whole plots of size k.

For example, a 3^2 factorial design is a $\mathcal{D}$-optimal 9-point design for estimating a full second order model in two variables when the errors are uncorrelated and possess a homogeneous variance. Arranging the nine runs of this design in a crossed design with three whole plots of size three therefore gives us a design that is $\mathcal{D}$-optimal in the class of crossed split-plot designs with three whole plots of size three for estimating a full second order model in one whole plot variable and one sub-plot variable.

It is a well-known result that 2^m designs and 2^{m-f} designs are $\mathcal{D}$-optimal for model (6.8). Since the points of 2^m designs can easily be arranged in crossed split-plot designs with $b = 2^{m_w}$ whole plots of size $k = 2^{m_s}$, the resulting designs are $\mathcal{D}$-optimal for estimating a first order model with or without interactions in the class of crossed split-plot designs . The 2^{m-f} designs can also be arranged in crossed split-plot designs. For this purpose, no sub-plot factors should be confounded with whole plot factors. The resulting designs are also $\mathcal{D}$-optimal in the class of crossed split-plot designs.

6.5.2 $\mathcal{D}$-optimality of 2^m, 2^{m-f} and Plackett-Burman split-plot designs

For a main effects model with uncorrelated errors and homogeneous variance, the $\mathcal{D}$-optimal factor levels are -1 and +1 if the region of interest is $[-1, +1]^m$. Box and Draper (1971) prove this for saturated designs, that is designs where $n = p$, using the fact that $|\mathbf{X}'\mathbf{X}| = |\mathbf{X}|^2$. As the $\mathcal{D}$-optimality criterion $|\mathbf{X}'\mathbf{V}^{-1}\mathbf{X}|$ for a model with different error assumptions can be written as $|\mathbf{V}^{-1}||\mathbf{X}|^2$ for saturated designs, the proof can be extended to models with correlated errors if $\mathbf{V}$ is identical for all design options available. For designs where $n > p$, the optimality of the levels -1 and +1 is shown in Mitchell (1974b). As a consequence of equation (6.7), this result is valid for crossed split-plot designs as well. These results are valid for models involving interactions as well because these models can be seen as restricted versions of main effects models in which the levels of some factors are the product of at least two other factors.

It also turns out that 2^m, 2^{m-f} and Plackett-Burman designs are $\mathcal{D}$-optimal among all feasible split-plot designs with b whole plots of size k, provided the levels of the sub-plot variables sum to zero within each whole plot. As a matter of fact, arranging the points of a $\mathcal{D}$-optimal design for the uncorrelated model (6.8) in b whole plots of size k so that the levels of the sub-plot factors sum to zero within each whole plot yields a $\mathcal{D}$-optimal split-plot design. This is proven in Appendix B.

The condition that the levels of the sub-plot factors sum to zero is satisfied for all 2^m split-plot designs and for all 2^{m-f} designs, the sub-plot generators of which contain at least two sub-plot factors. This is the case for all designs listed in Huang et al. (1998) and Bingham and Sitter (1999, 2001). The proof in Appendix B is valid for any design that maximizes $|\mathbf{X}'\mathbf{X}|$ for a given design region and whose design points can be arranged in whole plots so that the levels of the sub-plot variables sum to zero within each whole plot. Plackett-Burman designs can therefore be used to construct $\mathcal{D}$-optimal split-plot experiments in some instances. Finally, it is interesting to note that the equivalency between OLS and GLS is maintained for these type of split-plot designs.

In view of the popularity of two-level fractional factorial split-plot designs, this theoretical result is important to practitioners. Therefore, 2^m and 2^{m-f} split-plot designs receive detailed attention in Section 9.3. Nevertheless, it should be stressed that many experimental situations exist in which these kind of designs cannot provide the experimenter with a $\mathcal{D}$-optimal design. In these situations, a design construction algorithm is needed.

6.6   Design construction algorithm

The exchange algorithm we have developed for designing constrained split-plot experiments to some extent resembles that for computing optimal designs for the random block effects model. This should be no surprise in view of the strong resemblance between both design problems. Nevertheless, some important differences exist. In the description of the algorithm, we will denote a whole plot factor level by $\mathbf{w}_i$ and a candidate point or a design point by $(\mathbf{w}_i, \mathbf{s}_{ij})$.

The algorithm first computes a feasible n-point starting design. Firstly, a whole plot factor level $\mathbf{w}_i$ is randomly assigned to each of the whole plots, making sure that there are at least as many different levels as there are pure whole plot factor coefficients in the model, that is p_w. This restriction is necessary to avoid ending up with a singular design. Next, a random number of points $(\mathbf{w}_i, \mathbf{s}_{ij})$ is selected at random from the list of candidate design points. Each of these points is then randomly assigned to one of the whole plots with whole plot factor level $\mathbf{w}_i$. Finally, the starting design is completed by sequentially adding the candidate point with the largest prediction variance to the design. When a point $(\mathbf{w}_i, \mathbf{s}_{ij})$ cannot be added because all whole plots with whole plot factor level $\mathbf{w}_i$ are complete, the design point with the next largest prediction variance is added instead.

Next, three exchange strategies are considered to improve this initial design. Two of these strategies were already used in the previous chapter, namely the substitution of a design point with a candidate point and the interchange of two design points from different whole plots. The third strategy considered is the substitution of a whole plot factor level. The value of this strategy is that it adds flexibility to the algorithm.

Firstly, the substitution of a design point with a point of the candidate list is considered. This strategy was already used in the algorithm of Chapter 4. However, only a limited number of exchanges need to be evaluated for the design problem at hand. This is because a design point $(\mathbf{w}_i, \mathbf{s}_{ij})$ can only be replaced by a point with whole plot factor level $\mathbf{w}_i$, that is by a point of the form $(\mathbf{w}_i, \mathbf{s}_{ik})$, where $k \neq j$.

A second strategy considered to improve the design is the interchange of design points from different whole plots. The interchange of design points from different blocks was already used in the algorithm of Chapter 4 for the design of blocked experiments. However, the number of possible interchanges in the case of split-plot designs is limited because only points from whole plots with the same whole plot factor level can be interchanged.

In the third strategy, the substitution of whole plot factor levels is evaluated. This strategy differs from the two others in that all design points of a given whole plot are modified by the substitution. As a result, the evaluation of this type of exchanges is computationally much more prohibitive than the two other strategies. Suppose that we would like to replace the whole plot factor level $\mathbf{w}_i$ of the ith whole plot by $\mathbf{w}_i^*$. The information matrix $\mathbf{M}$ can then be updated by subtracting

$$\sum_{j=1}^{k_i} \mathbf{f}(\mathbf{w}_i, \mathbf{s}_{ij})\mathbf{f}'(\mathbf{w}_i, \mathbf{s}_{ij}) - \frac{\eta}{1 + k_i\eta}(\mathbf{X}_i'\mathbf{1}_{k_i})(\mathbf{X}_i'\mathbf{1}_{k_i})',$$

and adding

$$\sum_{j=1}^{k_i} \mathbf{f}(\mathbf{w}_i^*, \mathbf{s}_{ij})\mathbf{f}'(\mathbf{w}_i^*, \mathbf{s}_{ij}) - \frac{\eta}{1 + k_i\eta}(\mathbf{X}_i^{*'}\mathbf{1}_{k_i})(\mathbf{X}_i^{*'}\mathbf{1}_{k_i})',$$

where $\mathbf{X}_i$ and $\mathbf{X}_i^*$ are the parts of the design matrix corresponding to the ith whole plot before and after the change respectively. Fortunately, not all the elements of $\mathbf{M}$ are affected by the design change. This is due to the fact that only the whole plot factor levels are modified. As a result, the elements of the information matrix corresponding to the intercept, the sub-plot factors and the sub-plot by sub-plot interactions remain unchanged. Suppose that there are p_s coefficients in the model that do not involve whole plot factors, then p_s^2 elements of $\mathbf{M}$ do not require an update. This exchange strategy is extremely important for the algorithm as it is the only strategy that is able to change the whole plot factor level of a whole plot.

In the algorithm, more than one try is used in order to increase the probability of finding the optimal design. The input to the algorithm consists of the number of observations n, the number of whole plots b, the whole plot sizes k_i ($i = 1, 2, \ldots, b$), the number of model parameters p, the order of the model, the number of explanatory variables m, the structure of their polynomial expansion and the number of tries t. In addition, the whole plot and the sub-plot factors must be identified. Finally, an estimate of η must be provided. Information on η is typically available from the literature or from prior experiments of a similar kind. It turns out that degrees of correlation exceeding unity are no exception in split-plot experiments. For example, Letsinger et al. (1996) obtain $\hat{\eta} = 1.04$ for the chemical experiment described in Example 1 of Chapter 3. According to Bisgaard and Steinberg (1997), the whole plot error variance is usually larger than the sub-plot error variance. In other words, in many cases $\eta \geq 1$. In some cases, however, η will be smaller than one. In the vinyl thickness experiment analyzed in Section 3.3.2, for example, $\hat{\eta} = 0.82$. Depending on how they were analyzed, prior split-plot experiments may directly or indirectly contain valuable information on the degree of correlation in a specific experimental setting. If they were properly analyzed as a split-plot experiment, prior guesses for η are obtained by $\hat{\sigma}_\gamma^2 / \hat{\sigma}_\epsilon^2$, where $\hat{\sigma}_\epsilon^2$ and $\hat{\sigma}_\gamma^2$ are the estimates of the sub-plot and whole plot error variance respectively. If they were improperly analyzed as a completely randomized experiment, the data from the experiments can be recovered in order to analyze them properly and thereby obtain estimates of σ_ϵ^2 and σ_γ^2. Ganju and Lucas (1999) state that split-plot experiments are often erroneously analyzed as a completely randomized experiment. They also point out that information about the run order of the experiment and about the randomization is usually nonexistent, even though it is indispensable for a correct analysis. In this way, valuable information to construct prior guesses of η is lost as well. Fortunately, for the purpose of the construction of three-level second order designs, it turns out that a reasonable guess of η is satisfactory.

By default, the algorithm computes the grid of candidate points $G = \{1, 2, \ldots, g\}$ as in Atkinson and Donev (1992): the design region is assumed to be hypercubic and is taken as $[-1, +1]^m$. The grid points are chosen from the 2^m, 3^m, 4^m, $\ldots$ factorial design depending on whether the model contains linear, quadratic, cubic or higher order terms. Alternatively, the user can specify G if another set of candidate points is desired. This is important when the design region is hyperspherical, when it is restricted, or when the experimenter would like to search over a finer grid. In order to find efficient designs, the set of candidate points should certainly include corner points and cover the entire design region. For example, the points of a 3^m factorial design are not sufficient to find an efficient design to estimate a quadratic model on a hyperspherical region. In that case, star points should be included in the set of candidate points. Finally, note that the construction of a non-singular design requires that $n \geq p$, $g \geq p$ and that the number of different whole plots is greater than or equal to the number of pure whole plot coefficients p_w. Further details on the algorithm are given in Appendix C.

6.7   Computational results

We have used the algorithm for several combinations of b, k_i and n. A selection of the computational results is displayed in this section. Firstly, we have investigated the $\mathcal{D}$-optimal three-level designs for a full quadratic model obtained from a search over a coarse grid of candidate points. As usual, this coarse grid consisted of the 3^m factorial design. Next, we have performed a number of searches over an 11^m factorial design.

6.7.1   Coarse grid

Using the algorithm outlined in the previous section, we have performed a search over the points of the 3^3 factorial design in order to find the 18-point $\mathcal{D}$-optimal three-level design for a full quadratic model in one whole plot factor and two sub-plot factors. It was assumed that six whole plots of size three were available. The three split-plot designs we obtained are denoted by SPD1, SPD2 and SPD3, and they are displayed in Table 6.3. All three designs have three whole plots with whole plot factor level -1, one whole plot with whole plot factor level 0 and two whole plots with whole plot factor level +1. Reversing the signs of the whole plot factor levels yields designs that are equivalent to the original. In terms of $\mathcal{D}$-efficiency, SPD1 is the best possible three-level split-plot design with six whole plots of size three when $\eta \leq 0.2188$. When η lies between 0.2188 and 0.3787, SPD2 is the best design option, and for $\eta \geq 0.3787$, SPD3 is $\mathcal{D}$-optimal.

In order to visualize the performance of these three design options, we have computed the relative $\mathcal{D}$-efficiencies

$$\left\{ \frac{|\mathbf{X}'\mathbf{V}^{-1}\mathbf{X}|}{|\mathbf{A}'\mathbf{V}^{-1}\mathbf{A}|} \right\}^{1/p}, \tag{6.9}$$

where $\mathbf{X}$ is the design matrix of the split-plot design under consideration, and $\mathbf{A}$ is the design matrix of a benchmark design, for values of η between 0 and 10. The benchmark design contains two whole plots at the whole plot factor levels -1, 0 and +1. For the sub-plot factor levels, a duplicated 3^2 design was used. The runs of this design were assigned to the whole plot levels using the algorithm of Trinca and Gilmour (2001). The resulting design is also displayed in Table 6.3.

Table 6.3: $\mathcal{D}$-optimal split-plot designs and benchmark design with six whole plots of size three for the estimation of a full quadratic model in one whole plot factor and two sub-plot factors.

w	SPD1 s_1	SPD1 s_2	SPD2 s_1	SPD2 s_2	SPD3 s_1	SPD3 s_2	Bench w	Bench s_1	Bench s_2
-1	0	-1	-1	1	-1	0	-1	1	1
-1	-1	1	1	0	0	-1	-1	0	-1
-1	1	0	-1	-1	1	1	-1	-1	0
-1	-1	-1	-1	1	1	-1	-1	1	0
-1	1	1	1	-1	0	1	-1	0	-1
-1	0	0	0	1	-1	0	-1	-1	-1
-1	1	-1	-1	1	-1	1	0	1	1
-1	-1	-1	0	-1	-1	0	0	1	-1
-1	-1	-1	-1	1	1	-1	0	0	0
0	0	1	0	1	0	0	0	0	0
0	-1	0	1	-1	-1	-1	0	1	1
0	1	-1	-1	0	1	1	0	-1	1
1	1	1	0	0	-1	-1	1	1	1
1	-1	-1	-1	-1	0	1	1	1	-1
1	1	1	1	1	0	0	1	-1	0
1	-1	1	1	1	0	0	1	0	-1
1	1	0	1	1	1	0	1	-1	1
1	0	-1	-1	1	-1	-1	1	1	0

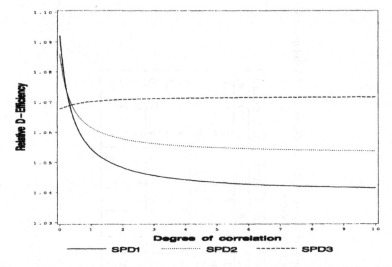

Figure 6.2: Relative $\mathcal{D}$-efficiencies of the three split-plot designs in Table 6.3 with respect to the benchmark design.

The relative $\mathcal{D}$-efficiencies of SPD1, SPD2 and SPD3 are displayed in Figure 6.2. The figure clearly shows that, for any value of η, all three split-plot designs are well over 4% more efficient than the benchmark design. SPD3 is 7% more efficient than the benchmark design for any value of η. The figure also visualizes that SPD1 is the best choice for small degrees of correlation. However, as η grows larger, SPD1 is overtaken by SPD2. When $\eta \geq 0.3787$, SPD3 becomes the best design option. The figure also shows that the efficiency of SPD1 and SPD2 is not very different for small degrees of correlation. However, SPD3 becomes substantially better than SPD1 and SPD2 when η approaches unity. The efficiency gain of using SPD3 instead of SPD1 is between 2% and 3% when $\eta \geq 1$.

Comparing the relative $\mathcal{D}$-efficiency of SPD1, SPD2 and SPD3 when $\eta = 0$ shows that using the design points of SPD1 in a completely randomized design would result in an efficient design. In other words, the projection of SPD1 obtained by ignoring the whole plots yields an efficient design for the uncorrelated model (6.8). Using the points of SPD2 would result in a completely randomized design that is slightly less efficient and using the points of SPD3 would result in a design that is less than 98% as efficient as SPD1. As a result, the optimal split-plot design for a small value of η is obtained from an efficient design for the uncorrelated model (6.8). When η is large, the optimal split-plot design can be obtained from an inefficient design for the uncorrelated model.

It is important to stress that the designs are optimal for wide ranges of η, especially if η is not too close to zero. In practice, it is unlikely that η is very small (see, for instance, Bisgaard and Steinberg 1997 and Bingham and Sitter 2001). As a consequence, precise knowledge of η is not required to construct $\mathcal{D}$-optimal split-plot designs. Similar results were obtained for other models, for other number of whole plots and for other whole plot sizes.

6.7.2   Fine grid

In some cases, using a finer grid over the experimental region than the points of the 3^m factorial design leads to a substantially better design. When the interest is in the estimation of a full quadratic model in one whole plot variable and one sub-plot variable and when five whole plots of size two are available, then gains in $\mathcal{D}$-efficiency of up to 1.5% can be achieved by using a 11^2 instead of a 3^2 grid. On the contrary, only gains of up to 0.3% can be realized when the number of whole plots available is four. This example shows that the gains heavily depend on all parameters of the design problem and that it is hard to predict when the search over a finer grid will produce substantially better designs. An alternative to a search over a finer grid is to adapt the adjustment algorithm of Donev and Atkinson (1988).

6.8   The protein extraction experiment

Let us reconsider the protein experiment in which 42 experimental runs must be assigned to 21 whole plots of size two. The experiment was already introduced in Example 2 of Chapter 3. A three-level design for this experiment is suggested by Trinca and Gilmour (2001). They use seven whole plots at each of the three whole plot factor levels -1, 0 and 1. For the sub-plot factors, they suggest using a central composite design with two center points and duplicated axial points. The factorial portion of the central composite design consists of the complete 2^4 factorial design plus half a fraction of it. The 42 runs of this design are displayed in Table 6.4. For this design problem, we have computed the $\mathcal{D}$-optimal three-level design for several values of η. The design displayed in Table 6.4 is $\mathcal{D}$-optimal for η-values from 1 to 10.

The $\mathcal{D}$-optimal design has nine whole plots with whole plot factor level -1, three whole plots with level 0 and nine with level +1. The $\mathcal{D}$-optimality criterion value of this design is of course larger than that of the design proposed by Trinca and Gilmour (2001). This is no surprise because the points of the latter design were not selected according to the $\mathcal{D}$-optimality

Table 6.4: Two experimental designs for the protein experiment.

Whole Plot	Trinca & Gilmour					$\mathcal{D}$-optimal design				
	w	s_1	s_2	s_3	s_4	w	s_1	s_2	s_3	s_4
1	-1	-1	-1	-1	-1	-1	1	-1	-1	1
	-1	1	-1	-1	1	-1	1	1	1	-1
2	-1	-1	1	-1	1	-1	-1	0	-1	-1
	-1	0	0	0	0	-1	1	1	-1	1
3	-1	-1	0	0	0	-1	-1	1	1	1
	-1	1	1	-1	-1	-1	1	1	-1	-1
4	-1	-1	1	1	-1	-1	-1	-1	1	1
	-1	1	1	-1	1	-1	1	-1	-1	-1
5	-1	0	0	0	-1	-1	-1	-1	1	-1
	-1	0	1	0	0	-1	0	1	0	0
6	-1	0	0	0	0	-1	-1	1	-1	1
	-1	1	-1	1	-1	-1	1	-1	1	0
7	-1	0	-1	0	0	-1	-1	-1	-1	1
	-1	1	1	1	1	-1	-1	1	1	-1
8	0	-1	-1	-1	1	-1	-1	1	-1	-1
	0	1	-1	1	-1	-1	1	-1	1	-1
9	0	-1	-1	1	-1	-1	-1	-1	0	0
	0	1	-1	1	1	-1	1	0	1	1
10	0	-1	-1	1	1	0	0	0	-1	0
	0	1	-1	-1	-1	0	1	-1	0	1
11	0	-1	1	-1	-1	0	0	-1	0	-1
	0	1	1	1	1	0	1	0	-1	0
12	0	-1	1	1	1	0	0	-1	1	1
	0	1	1	1	-1	0	1	1	1	-1
13	0	-1	0	0	0	1	-1	-1	1	-1
	0	0	0	0	1	1	1	1	-1	-1
14	0	0	0	1	0	1	-1	0	0	1
	0	1	0	0	0	1	1	1	1	1
15	1	-1	-1	-1	-1	1	-1	-1	-1	-1
	1	1	0	0	0	1	0	1	1	-1
16	1	-1	-1	1	1	1	-1	-1	1	1
	1	0	1	0	0	1	1	1	-1	1
17	1	-1	1	1	-1	1	0	-1	-1	1
	1	0	0	-1	0	1	1	-1	1	0
18	1	-1	1	-1	1	1	-1	1	-1	-1
	1	1	1	-1	-1	1	1	0	0	1
19	1	0	0	-1	0	1	-1	0	0	-1
	1	0	-1	0	0	1	1	-1	-1	1
20	1	1	-1	-1	1	1	-1	1	1	1
	1	0	0	0	-1	1	1	-1	1	-1
21	1	0	0	0	1	1	-1	1	-1	1
	1	0	0	1	0	1	1	-1	-1	-1

Table 6.5: Expected variances for the parameter estimates of a second order model from both split-plot designs given in Table 6.4.

	Trinca & Gilmour			$\mathcal{D}$-optimal design		
	$\eta = 1$	$\eta = 5$	$\eta = 10$	$\eta = 1$	$\eta = 5$	$\eta = 10$
w	0.118	0.407	0.765	0.086	0.309	0.586
s_1	0.041	0.044	0.045	0.036	0.039	0.039
s_2	0.077	0.124	0.140	0.038	0.042	0.043
s_3	0.056	0.065	0.068	0.039	0.044	0.045
s_4	0.045	0.050	0.051	0.040	0.048	0.049
ws_1	0.097	0.113	0.118	0.036	0.038	0.039
ws_2	0.115	0.182	0.206	0.041	0.046	0.047
ws_3	0.118	0.164	0.179	0.041	0.047	0.048
ws_4	0.095	0.114	0.120	0.044	0.053	0.056
s_1s_2	0.056	0.060	0.061	0.056	0.077	0.083
s_1s_3	0.074	0.091	0.095	0.055	0.075	0.082
s_1s_4	0.123	0.262	0.331	0.051	0.074	0.083
s_2s_3	0.081	0.114	0.125	0.056	0.079	0.087
s_2s_4	0.053	0.054	0.055	0.052	0.071	0.078
s_3s_4	0.073	0.096	0.102	0.048	0.061	0.066
w^2	0.384	1.260	2.338	0.683	2.261	4.213
s_1^2	0.316	0.393	0.414	0.322	0.356	0.364
s_2^2	0.305	0.390	0.415	0.264	0.293	0.301
s_3^2	0.281	0.337	0.352	0.260	0.304	0.316
s_4^2	0.281	0.337	0.352	0.308	0.357	0.371

criterion. The relative $\mathcal{D}$-efficiency of Trinca and Gilmour's design with respect to the $\mathcal{D}$-optimal design amounts to 76.77%, 73.20% and 72.21% when η is 1, 5 or 10 respectively. The expected variances for the factor effect estimates for both designs are given in Table 6.5.

It turns out that 16 of the 20 factor effects are estimated more efficiently with the $\mathcal{D}$-optimal design when $\eta = 1$, $\eta = 5$ or $\eta = 10$. The effects that are estimated less efficiently by the $\mathcal{D}$-optimal design correspond to w^2, s_1^2 and s_4^2. From Table 6.5, it can also be seen that the variances obtained by using the $\mathcal{D}$-optimal design do not increase very much with η. As a result, the parameter estimates are not largely affected by the variation between the days of experimentation. This is because the sub-plots are nearly orthogonal to the whole plots. The $\mathcal{D}$-optimality criterion thus implicitly arranges the sub-plots in a near-orthogonal way, just like the algorithm of Trinca and Gilmour (2001).

If we compare the average variance of the factor effect estimates of both designs, we see that the $\mathcal{D}$-optimal design is 9.11% better when $\eta = 1$, nearly equivalent when $\eta = 5$ and 9.47% less efficient when $\eta = 10$. However, if we compare the average prediction variance of both designs, we see that the $\mathcal{D}$-optimal design is 16.67% more efficient when $\eta = 1$ and 7.51% more efficient when $\eta = 5$. When $\eta = 10$, the $\mathcal{D}$-optimal design is 0.73% less efficient than the design proposed by Trinca and Gilmour (2001). As a conclusion, the $\mathcal{D}$-optimal design is an excellent choice when the estimated model is used for prediction purposes, but also with respect to other criteria, such as orthogonality.

6.9   Algorithm evaluation

In order to evaluate the quality of our algorithm, we have performed 1,000 tries for computing optimal designs in several experimental situations. For each of the problems investigated, we have used 8 values of the degree of correlation η: 0.1, 0.25, 0.5, 0.75, 1, 2, 5, 10. In each case, we have estimated the probability of finding the best design as well as the expected efficiencies of the best design obtained from different numbers of tries. The computation of these expected efficiencies is given in Appendix D. It turns out that for simple problems, the optimal design is found in 100% of the tries. For harder problems, the probability of finding the optimum is excellent, as well as the estimated expected efficiency of the best design found.

Design problem 1

The first design problem considered is the problem of finding the best possible split-plot design with three whole plots of size three for the estimation of a full quadratic model in one whole plot variable and one sub-plot variable. The optimal design for this problem is the crossed design given in Figure 3.2. For any degree of correlation η used, the optimal design was found in every try.

Design problem 2

The second design problem considered is the problem of finding the 2^{6-1} fractional factorial split-plot design with three whole plot factors A, B and C, and three sub-plot factors D, E and F, where the factor F is confounded with the five factor interaction ABCDE (for more details on 2^m and 2^{m-f} split-plot designs, we refer the reader to Section 9.3). This split-plot design is a non-crossed design with eight whole plots of size four. It is $\mathcal{D}$-optimal for any value of η. It turns out that the algorithm is able to find this design in at least 64.8% of the tries.

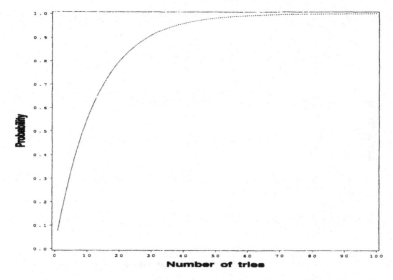

Figure 6.3: Estimated probability of finding the optimal design for Design Problem 3 when $\eta = 0.25$.

Design problem 3

The third problem considered was the construction of an experiment with six whole plots of size three for the estimation of a full quadratic model in one whole plot factor and two sub-plot factors. For most values of η, the estimated probability of finding the optimal design in one try lies between 0.200 and 0.450, which is excellent. However, for values of η around 0.5, the probability of finding the optimal design is smaller. For example, for $\eta = 0.25$, it is only 0.077. At first sight, this number is not very convincing. However, the estimated probability of finding the optimum rapidly goes to one if the number of tries t is increased. This is shown in Figure 6.3. In addition, a glance at Figure 6.2 reveals that SPD2 and SPD3 are nearly equivalent in the neighborhood of $\eta = 0.5$. Therefore, missing the optimum does not necessarily lead to a considerable loss in $\mathcal{D}$-efficiency. Whether this is the case or not can be seen from the estimated expected $\mathcal{D}$-efficiencies displayed in Figure 6.4. The figure shows that the estimated expected efficiency quickly increases to one. It is thus very likely that the algorithm will produce a highly efficient design.

Design problem 4

The final design problem considered was the construction of a constrained split-plot experiment with four variables, two of which were a whole plot variable. The number of whole plots was equal to seven and the number of

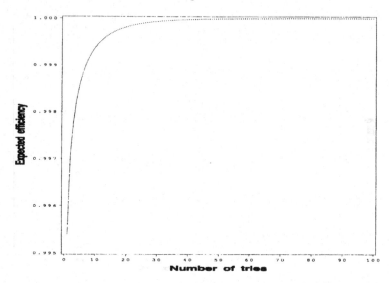

Figure 6.4: Estimated expected $\mathcal{D}$-efficiency of the best design found for Design Problem 3 when $\eta = 0.25$.

sub-plots within each whole plot was equal to four. A similar picture was obtained as for Design Problem 3: the probability of finding the optimum is small for small values of η because several designs are close to $\mathcal{D}$-optimal for small degrees of correlation. For larger values, the probability of finding the optimum is greater than 0.200.

6.10   Cost efficiency and statistical efficiency

The design problems considered in this chapter and in the previous chapters are very similar. The only difference is that, for the constrained split-plot experiment, it is required that all runs within one group (whole plot) possess the same level for the whole plot factors. This split-plot constraint is inspired by cost considerations and makes experiments with hard-to-change factors much easier to execute. The drawback is of course that the statistical efficiency suffers greatly from this constraint. In order to see this, let us compare the $\mathcal{D}$-criterion value of the best possible split-plot designs with one and with two whole plot factors to that of the $\mathcal{D}$-optimal blocked design. Assume that the interest is in estimating a full quadratic model in three variables and that six groups of three observations are available.

When η is as small as 0.1, the $\mathcal{D}$-criterion value of the $\mathcal{D}$-optimal blocked design amounts to 2.206E+9. When there is one whole plot variable, the $\mathcal{D}$-

criterion value drops to 1.431E+9 yielding a relative $\mathcal{D}$-efficiency of 95.8%. When there are two whole plot variables, the $\mathcal{D}$-criterion value is only 4.680E+8 yielding a relative $\mathcal{D}$-efficiency of 85.6%. This shows that, in comparison with a blocked design, a substantial amount of information is lost due to split-plotting. The loss of information is larger when two whole plot factors are used. When $\eta = 1$, the relative $\mathcal{D}$-efficiency of a split-plot design with one whole plot factor is 80.7%. For a split-plot design with two whole plot factors, it is only 52.9%. Finally, when $\eta = 10$, the relative $\mathcal{D}$-efficiencies of a split-plot design with one and two whole plot factors are 55.3% and 19.7% respectively. These results show that the loss of information increases dramatically with η.

The practical relevance of these figures lies in the fact that the experimenter often has to decide how many experimental variables will be treated as whole plot factors. Of course, the more variables will be treated as whole plot factors, the more time and costs can be saved during the phase of experimentation. However, these savings go at the expense of a substantial loss of information. In view of the utmost importance of the design phase in the development of new products and processes, the loss of information incurred by using many whole plot variables could have severe consequences in the long run.

Appendix A. Optimality of crossed split-plot designs

Assume that a design matrix $\mathbf{X}$ containing n experimental runs is given and that the observations should be assigned to b whole plots of size k. Denote the number of different whole plot factor levels by n_w and the number of whole plots with the ith whole plot factor level by b_i. In the example from Section 6.5.1, the number of whole plots b is equal to six, the whole plot size k is equal to three, the number of whole plot factor levels n_w is equal to three and the number of whole plots per whole plot factor level is two. Hence, $b_1 = b_2 = b_3 = 2$. Now, any assignment of the experimental runs to the whole plots divides the $n \times p$ dimensional design matrix $\mathbf{X}$ in b distinct $k \times p$ dimensional submatrices $\mathbf{X}_{ij}$. Each $\mathbf{X}_{ij}$ corresponds to the part of $\mathbf{X}$ that belongs to the jth whole plot with the ith whole plot factor level. The information matrix (3.15) can then be written as

$$\mathbf{M} = \frac{1}{\sigma_\varepsilon^2} \Big\{ \mathbf{X}'\mathbf{X} - \frac{\eta}{1+k\eta} \sum_{i=1}^{n_w} \sum_{j=1}^{b_i} (\mathbf{X}_{ij}'\mathbf{1}_k)(\mathbf{X}_{ij}'\mathbf{1}_k)' \Big\}. \qquad (6.10)$$

Now, denote by $\mathbf{W}_i$ the $k \times (p_w + 1)$ dimensional part of $\mathbf{X}_{ij}$ corresponding to the intercept and the p_w pure whole plot terms of the statistical model, and by $\mathbf{S}_{ij}$ the $k \times (p - p_w - 1)$ dimensional part of $\mathbf{X}_{ij}$ corresponding to the sub-plot terms and the whole plot by sub-plot interactions. Substituting

$\mathbf{X}_{ij} = [\, \mathbf{W}_i \quad \mathbf{S}_{ij}\,]$ and $c = \eta/(1 + k\eta)$ in (6.10) yields

$$\mathbf{M} = \frac{1}{\sigma_\varepsilon^2}(\mathbf{X'X} - c\mathbf{A}), \tag{6.11}$$

where

$$\mathbf{A} = \sum_{i=1}^{n_w} \sum_{j=1}^{b_i} \begin{bmatrix} (\mathbf{W}_i'\mathbf{1}_k)(\mathbf{W}_i'\mathbf{1}_k)' & (\mathbf{W}_i'\mathbf{1}_k)(\mathbf{S}_{ij}'\mathbf{1}_k)' \\ (\mathbf{S}_{ij}'\mathbf{1}_k)(\mathbf{W}_i'\mathbf{1}_k)' & (\mathbf{S}_{ij}'\mathbf{1}_k)(\mathbf{S}_{ij}'\mathbf{1}_k)' \end{bmatrix}.$$

In a crossed split-plot design, the levels of the sub-plot factors are the same in every whole plot. As a result, $\mathbf{S}_{ij}$ is identical for all whole plots with the ith whole plot factor level (if no whole plot by sub-plot interactions are present, then $\mathbf{S}_{ij}$ is even identical for all whole plots). Therefore, we can write $\mathbf{S}_{ij} = \mathbf{S}_i$ ($i = 1, 2, \ldots, n_w; j = 1, 2, \ldots, b_i$). As a consequence, $\mathbf{S}_{ij}'\mathbf{1}_k = \mathbf{S}_i'\mathbf{1}_k$ and the information matrix (6.11) can be written as

$$\mathbf{M}_{\text{crossed}} = \frac{1}{\sigma_\varepsilon^2}(\mathbf{X'X} - c\mathbf{A}_{\text{crossed}}), \tag{6.12}$$

where

$$\mathbf{A}_{\text{crossed}} = \sum_{i=1}^{n_w} \sum_{j=1}^{b_i} \begin{bmatrix} (\mathbf{W}_i'\mathbf{1}_k)(\mathbf{W}_i'\mathbf{1}_k)' & (\mathbf{W}_i'\mathbf{1}_k)(\mathbf{S}_i'\mathbf{1}_k)' \\ (\mathbf{S}_i'\mathbf{1}_k)(\mathbf{W}_i'\mathbf{1}_k)' & (\mathbf{S}_i'\mathbf{1}_k)(\mathbf{S}_i'\mathbf{1}_k)' \end{bmatrix}.$$

In a non-crossed design, the levels of the sub-plot factors are not identical in all whole plots and we have that

$$\mathbf{S}_{ij}'\mathbf{1}_k = \mathbf{S}_i'\mathbf{1}_k + \delta_{ij}, \quad (i = 1, 2, \ldots, n_w; j = 1, 2, \ldots, b_i),$$

where

$$\sum_{j=1}^{b_i} \delta_{ij} = \mathbf{0}_{p-p_w-1}, \quad (i = 1, 2, \ldots, n_w).$$

Substituting these equations in (6.11) yields

$$\mathbf{M}_{\text{n.crossed}} = \frac{1}{\sigma_\varepsilon^2}(\mathbf{X'X} - c\mathbf{A}_{\text{n.crossed}}), \tag{6.13}$$

where

$$\mathbf{A}_{\text{n.crossed}} = \sum_{i=1}^{n_w}\sum_{j=1}^{b_i}\begin{bmatrix}(\mathbf{W}_i'1_k)(\mathbf{W}_i'1_k)' & (\mathbf{W}_i'1_k)(\mathbf{S}_i'1_k + \boldsymbol{\delta}_{ij})' \\ (\mathbf{S}_i'1_k + \boldsymbol{\delta}_{ij})(\mathbf{W}_i'1_k)' & (\mathbf{S}_i'1_k + \boldsymbol{\delta}_{ij})(\mathbf{S}_i'1_k + \boldsymbol{\delta}_{ij})'\end{bmatrix},$$

$$= \mathbf{A}_{\text{crossed}} + \sum_{i=1}^{n_w}\sum_{j=1}^{b_i}\begin{bmatrix}\mathbf{0}_{(p_w+1)\times(p_w+1)} & (\mathbf{W}_i'1_k)\boldsymbol{\delta}_{ij}' \\ \boldsymbol{\delta}_{ij}(\mathbf{W}_i'1_k)' & \boldsymbol{\delta}_{ij}(\mathbf{S}_i'1_k)' + (\mathbf{S}_i'1_k)\boldsymbol{\delta}_{ij}' + \boldsymbol{\delta}_{ij}\boldsymbol{\delta}_{ij}'\end{bmatrix},$$

$$= \mathbf{A}_{\text{crossed}} + \sum_{i=1}^{n_w}\begin{bmatrix}\mathbf{0}_{(p_w+1)\times(p_w+1)} & \mathbf{0}_{(p_w+1)\times(p-p_w-1)} \\ \mathbf{0}_{(p-p_w-1)\times(p_w+1)} & \sum_{j=1}^{b_i}\boldsymbol{\delta}_{ij}\boldsymbol{\delta}_{ij}'\end{bmatrix},$$

$$= \mathbf{A}_{\text{crossed}} + \sum_{i=1}^{n_w}\sum_{j=1}^{b_i}\begin{bmatrix}\mathbf{0}_{p_w+1} \\ \boldsymbol{\delta}_{ij}\end{bmatrix}\begin{bmatrix}\mathbf{0}_{p_w+1} \\ \boldsymbol{\delta}_{ij}\end{bmatrix}',$$

$$= \mathbf{A}_{\text{crossed}} + \boldsymbol{\Delta}'\boldsymbol{\Delta},$$

where $\boldsymbol{\Delta} =$ is the $b \times p$ dimensional matrix the rows of which are given by $[\, \mathbf{0}_{p_w+1}'\ \boldsymbol{\delta}_{ij}'\,]$. As a result, the difference $\mathbf{A}_{\text{n.crossed}} - \mathbf{A}_{\text{crossed}}$ is nonnegative definite, as well as the difference $\mathbf{M}_{\text{crossed}} - \mathbf{M}_{\text{n.crossed}}$. Assigning the observations to the whole plots so that the resulting split-plot design is crossed is therefore an optimal strategy for a given design matrix $\mathbf{X}$.

Appendix B. $\mathcal{D}$-optimality of 2^m and 2^{m-f} designs

In this appendix, we show that arranging the points of a $\mathcal{D}$-optimal design for the uncorrelated model (6.8) in a split-plot design with b whole plots of size k so that the levels of the sub-plot variables sum to zero within each whole plot yields a $\mathcal{D}$-optimal split-plot design with b whole plots of size k. Firstly, we show that arranging the sub-plot treatment in such a way is $\mathcal{D}$-optimal for given sets of whole plot treatments and sub-plot treatments. Given this result, it is easy to see that 2^m and 2^{m-f} designs provide the optimal set of treatments for the experiment.

If we denote by $\mathbf{W}_i$ the part of $\mathbf{X}_i$ corresponding to the intercept and the pure whole plot coefficients, and by $\mathbf{S}_i$ the part corresponding to the remaining $p-p_w-1$ coefficients, the information matrix (3.15) of a split-plot experiment with b whole plots of size k can be written as

$$\mathbf{M} = \frac{1}{\sigma_\varepsilon^2}\left\{\begin{bmatrix}\mathbf{W}'\mathbf{W} & \mathbf{W}'\mathbf{S} \\ \mathbf{S}'\mathbf{W} & \mathbf{S}'\mathbf{S}\end{bmatrix} - c\sum_{i=1}^{b}\begin{bmatrix}(\mathbf{W}_i'1_k)(\mathbf{W}_i'1_k)' & (\mathbf{W}_i'1_k)(\mathbf{S}_i'1_k)' \\ (\mathbf{S}_i'1_k)(\mathbf{W}_i'1_k)' & (\mathbf{S}_i'1_k)(\mathbf{S}_i'1_k)'\end{bmatrix}\right\},$$

where $\mathbf{X} = [\ \mathbf{W}\ \ \mathbf{S}\ ]$, and c is defined as in Appendix A. Using the fact that the k rows of each $\mathbf{W}_i$ are equal, this can be reduced to

$$\mathbf{M} = \frac{1}{\sigma_\varepsilon^2}\begin{bmatrix}(1 - ck)\mathbf{W}'\mathbf{W} & (1 - ck)\mathbf{W}'\mathbf{S} \\ (1 - ck)\mathbf{S}'\mathbf{W} & \mathbf{S}'\mathbf{S} - c\sum_{i=1}^{b}(\mathbf{S}_i'1_k)(\mathbf{S}_i'1_k)'\end{bmatrix}. \tag{6.14}$$

From Harville's (1997) Theorem 13.3.8, we have that

$$|\mathbf{M}| = |\mathbf{U}| \, |(1 - ck)\mathbf{W}'\mathbf{W} - (1 - ck)\mathbf{W}'\mathbf{S}\mathbf{U}^{-1}(1 - ck)\mathbf{S}'\mathbf{W}|,$$
$$= (1 - ck)^{p_w+1} \, |\mathbf{U}| \, |\mathbf{W}'\mathbf{W} - (1 - ck)\mathbf{W}'\mathbf{S}\mathbf{U}^{-1}\mathbf{S}'\mathbf{W}|,$$

where $\mathbf{U} = \mathbf{S}'\mathbf{S} - c\sum_{i=1}^{b}(\mathbf{S}_i'\mathbf{1}_k)(\mathbf{S}_i'\mathbf{1}_k)'$.

If the sub-plot treatments are assigned to the whole plots so that their levels sum to zero within each whole plot, $\mathbf{S}_i'\mathbf{1}_k = \mathbf{0}_k$ ($i = 1, 2, \ldots, b$), $\mathbf{W}'\mathbf{S} = \mathbf{0}_{(p_w+1)\times(p-p_w-1)}$, (6.14) reduces to

$$\mathbf{M}^* = \frac{1}{\sigma_\varepsilon^2} \begin{bmatrix} (1 - ck)\mathbf{W}'\mathbf{W} & \mathbf{0}_{(p_w+1)\times(p-p_w-1)} \\ \mathbf{0}_{(p-p_w-1)\times(p_w+1)} & \mathbf{S}'\mathbf{S} \end{bmatrix}, \qquad (6.15)$$

and

$$|\mathbf{M}^*| = |(1 - ck)\mathbf{W}'\mathbf{W}| \, |\mathbf{S}'\mathbf{S}| = (1 - ck)^{p_w+1} \, |\mathbf{W}'\mathbf{W}| \, |\mathbf{S}'\mathbf{S}|. \qquad (6.16)$$

From Harville's (1997) Corollary 18.1.8, we have that

$$|\mathbf{W}'\mathbf{W}| > |\mathbf{W}'\mathbf{W} - (1 - ck)\mathbf{W}'\mathbf{S}\mathbf{U}^{-1}\mathbf{S}'\mathbf{W}|$$

and

$$|\mathbf{S}'\mathbf{S}| > \left|\mathbf{S}'\mathbf{S} - c\sum_{i=1}^{b}(\mathbf{S}_i'\mathbf{1}_k)(\mathbf{S}_i'\mathbf{1}_k)'\right|.$$

As a result, $|\mathbf{M}^*| > |\mathbf{M}|$. For any set of sub-plot treatments $\mathbf{S}$, the best assignment in terms of $\mathcal{D}$-optimality is such that $\mathbf{S}_i'\mathbf{1}_k = \mathbf{0}_k$ ($i = 1, 2, \ldots, b$), no matter what the set of whole plot treatments $\mathbf{W}$ is.

It is well-known that 2^m, 2^{m-f} and Plackett-Burman designs maximize $|\mathbf{X}'\mathbf{X}| = |\mathbf{W}'\mathbf{W}| \, |\mathbf{S}'\mathbf{S}|$. For specific values of b and k, these designs can be arranged so that $\mathbf{S}_i'\mathbf{1}_k = \mathbf{0}_k$ ($i = 1, 2, \ldots, b$) and their determinant equals $(1 - ck)^{p_w+1}|\mathbf{X}'\mathbf{X}| = (1 - ck)^{p_w+1}n^p$. Any other design $\mathbf{X}^*$ for which $|\mathbf{X}^{*'}\mathbf{X}^*| < |\mathbf{X}'\mathbf{X}|$ leads to a determinant which cannot exceed $(1 - ck)^{p_w+1}|\mathbf{X}^{*'}\mathbf{X}^*|$. 2^m, 2^{m-f} and Plackett-Burman designs are therefore $\mathcal{D}$-optimal when used in a split-plot experiment with appropriate values of b and k.

Appendix C. The construction algorithm

We denote the set of g candidate points by G, the set of b whole plots by B, the set of k_i not necessarily distinct design points belonging to the ith whole plot of a given design D by D_i ($i = 1, 2, \ldots, b$), the whole plot factor level corresponding to the ith whole plot by $\mathbf{w}_i$, the set of candidate points with the ith whole plot factor level by G_i and the $\mathcal{D}$-criterion value of a given design D by $\mathcal{D}$. The best design found at a given time by the algorithm will be denoted by D^*. Its whole plots will be denoted by D_i^* $(i = 1, 2, \ldots, b)$ and the corresponding $\mathcal{D}$-criterion value by $\mathcal{D}^*$. For simplicity, we denote the information matrix of the experiment by $\mathbf{M}$. The singularity while constructing a starting design is overcome by using $\mathbf{M} + \omega\mathbf{I}$ instead of $\mathbf{M}$ with ω a small positive number. The set of possible whole plot factor levels is denoted by W. Finally, we denote the number of tries by t and the number of the current try by t_c. The algorithm starts by specifying the set of grid points $G = \{1, 2, \ldots, g\}$ and proceeds as follows:

1. Set $\mathcal{D}^* = 0$ and $t_c = 1$.

2. Determine the number of pure whole plot coefficients p_w.

3. Determine the number of pure sub-plot coefficients p_s.

4. Set $\mathbf{M} = \omega\mathbf{I}$ and $D_i = \emptyset$ $(i = 1, 2, \ldots, b)$.

5. Generate starting design.

   (a) Randomly assign p_w different whole plot factor levels $\mathbf{w}$ to p_w different whole plots.
   (b) Randomly assign $b - p_w$ whole plot factor levels to the remaining whole plots.
   (c) Randomly choose r $(1 \leq r \leq p)$.
   (d) Do r times:
      i. Randomly choose $i \in B$.
      ii. Randomly choose $j \in G_i$.
      iii. If $\#D_i < k_i$, then $D_i = D_i \cup \{j\}$, else go to step i.
      iv. Update $\mathbf{M}$.
   (e) Do $n - r$ times:
      i. Set $l = 1$.
      ii. Determine $j \in G$ with lth largest prediction variance.
      iii. Find i where $i \in B$, $j \in G_i$ and $\#D_i < k_i$. If no such i exists, set $l = l + 1$ and go to step ii.
      iv. $D_i = D_i \cup \{j\}$.
      v. Update $\mathbf{M}$.

6. Compute $\mathbf{M}$ and $\mathcal{D}$. If $\mathcal{D} = 0$, then go to step 4, else continue.

7. Set $\kappa = 0$.

8. Evaluate exchanges of design points.

   (a) Set $\delta = 1$.
   (b) $\forall i \in B, \forall j \in D_i, \forall k \in G_i, j \neq k$:
      i. Compute the effect $\delta_{jk}^i = \mathcal{D}'/\mathcal{D}$ of exchanging j by k in the ith whole plot.
      ii. If $\delta_{jk}^i > \delta$, then $\delta = \delta_{jk}^i$ and store i, j and k.

9. If $\delta > 1$, then go to step 10, else go to step 11.

10. Carry out best exchange.

    (a) $D_i = D_i\backslash\{j\} \cup \{k\}$.
    (b) Update $\mathbf{M}$ and $\mathcal{D}$.
    (c) Set $\kappa = 1$.

11. Evaluate interchanges.

    (a) Set $\delta = 1$.
    (b) $\forall i, j \in B, i < j, \mathbf{w}_i = \mathbf{w}_j, \forall k \in D_i, \forall l \in D_j, k \neq l$:

        i. Compute the effect $\delta_{ik}^{jl} = \mathcal{D}'/\mathcal{D}$ of moving k from whole plot i to j and l from whole plot j to i.

        ii. If $\delta_{ik}^{jl} > \delta$, then $\delta = \delta_{ik}^{jl}$ and store i, j, k and l.

12. If $\delta > 1$, then go to step 13, else go to step 14.

13. Carry out best interchange.

    (a) $D_i = D_i\backslash\{k\} \cup \{l\}$.
    (b) $D_j = D_j\backslash\{l\} \cup \{k\}$.
    (c) Update $\mathbf{M}$ and $\mathcal{D}$.
    (d) Set $\kappa = 1$.

14. Evaluate exchanges of whole plot factor levels.

    (a) Set $\delta = 1$.
    (b) $\forall i \in B, \forall j \in W, \mathbf{w}_i \neq \mathbf{w}_j$:

        i. Compute the effect $\delta_{ij} = \mathcal{D}'/\mathcal{D}$ of exchanging $\mathbf{w}_i$ by $\mathbf{w}_j$ in the ith whole plot.

        ii. If $\delta_{ij} > \delta$, then $\delta = \delta_{ij}$ and store i and j.

15. If $\delta > 1$, then go to step 16, else go to step 17.

16. Carry out best exchange.

    (a) Update D_i and G_i.
    (b) Update $\mathbf{M}$ and $\mathcal{D}$.
    (c) Set $\kappa = 1$.

17. If $\kappa = 1$, go to step 7.

18. If $\mathcal{D} > \mathcal{D}^*$, then $\mathcal{D}^* = \mathcal{D}, \forall i \in B : D_i^* = D_i$.

19. If $t_c < t$, then $t_c = t_c + 1$ and go to step 4, else stop.

In order to speed up the algorithm, the powerful update formulae and routines described in the Chapters 1, 2 and 3 are used during the construction of the starting design and the evaluation of the three exchange strategies. In order to save memory space, only the upper diagonal elements of the information matrix $\mathbf{M}$ are computed. This is possible because $\mathbf{M}$ is symmetric.

Appendix D. Estimated expected efficiency

Assume that for a large number of tries t, we obtain d distinct designs $D_1, D_2, \ldots, D_d$, with efficiencies $E_1 > E_2 > \cdots > E_d$. D_1 then is the best design and an estimate of the probability of finding D_1 in t tries, say p_1, is given by the number of times D_1 is found divided by t. In t tries, the number of times we find D_1 when the probability of success equals p_1 is binomially distributed with parameters t and p_1. The probability of finding the best design in t tries is $1-(1-p_1)^t$. The joint distribution of the number of times $F_1, F_2, \ldots, F_d$ we obtain $D_1, D_2, \ldots, D_d$ has a multinomial distribution with parameters t and $p_1, p_2, \ldots, p_d$. The probability that D_i is the best design found equals

$$P(\text{best design is } D_i) = P(F_i \geq 1 \text{ and } F_1 = F_2 = \cdots = F_{i-1} = 0),$$

$$= \sum_{k=1}^{t} P(F_i = k \text{ and } \sum_{j=1}^{i-1} F_j = 0 \text{ and } \sum_{j=i+1}^{d} F_j = t - k),$$

$$= \sum_{k=1}^{t} \binom{t}{k} p_i^k \left(\sum_{j=i+1}^{d} p_j \right)^0 \left(\sum_{j=i+1}^{d} p_j \right)^{t-k},$$

$$= \sum_{k=1}^{t} \binom{t}{k} p_i^k \left(\sum_{j=i+1}^{d} p_j \right)^{t-k},$$

$$= \sum_{k=0}^{t} \binom{t}{k} p_i^k \left(\sum_{j=i+1}^{d} p_j \right)^{t-k} - \left(\sum_{j=i+1}^{d} p_j \right)^t,$$

$$= \left(p_i + \sum_{j=i+1}^{d} p_j \right)^t - \left(\sum_{j=i+1}^{d} p_j \right)^t,$$

$$= \left(\sum_{j=i}^{d} p_j \right)^t - \left(\sum_{j=i+1}^{d} p_j \right)^t.$$

As a result, the expected efficiency from t tries can be written as

$$E(\text{efficiency}) = \sum_{i=1}^{d-1} \left[\left(\sum_{j=i}^{d} p_j \right)^t - \left(\sum_{j=i+1}^{d} p_j \right)^t \right] E_i + p_d^t E_d. \qquad (6.17)$$

7

Optimal Split-Plot Designs in the Presence of Hard-to-Change Factors

In this chapter, we will continue focusing on split-plot designs. However, we no longer assume that the number of whole plots and the size of the whole plots are dictated by the experimental situation. Instead, another type of restriction is imposed on the split-plot design: only one whole plot is connected to each combination of the hard-to-change or whole plot factor levels. A typical example of this sort of split-plot design is the prototype experiment, where exactly one prototype is built for each combination of the whole plot factor levels.

7.1 Introduction

As already pointed out in the previous chapter, conducting a completely randomized design (CRD) is impractical and can be highly inconvenient and very costly in cases where factor levels are difficult to change or to control. Typical examples of such factors are pressure, humidity and process temperature. Rather than conducting a CRD in which pressure has to be moved back and forth according to the randomization scheme, successively executing experimental runs with equal pressure will be preferred by the experimenter. Letsinger, Myers and Lentner (1996), Ganju and Lucas (1999) and Lucas and Ju (2002) describe how split-plot designs (SPDs) are obtained by not resetting the factor levels for the consecutive runs of an experiment.

As already pointed out in Example 1 of Chapter 3, Letsinger et al. (1996) describe an experiment from a U.S. chemical company in which the effect of five process variables, called temperature 1, temperature 2, humidity 1, humidity 2 and pressure, on a certain quality characteristic was investigated. A modified central composite design with 28 runs was used to conduct the experiment. The different factor level combinations of the design were not carried out in a completely random order because the levels of the factors temperature 1 and pressure were hard to change. Instead, all the runs with the same level for these two factors were grouped and all runs within one group were carried out immediately after each other. In doing so, it was much easier to conduct the experiment because the levels of the hard-to-change factors were changed as little as possible. Lucas and Ju (2002) call the resulting run order a completely restricted run order.

Bisgaard and Steinberg (1997) use an example from Taguchi (1989) to illustrate how prototype experiments are designed. The purpose of the experiment, which was already discussed on page 74, was to reduce the CO content of exhaust gas. Seven hard-to-change factors, A, B, C, D, E, F and G, each possessing two levels, were studied, along with three driving modes R_1, R_2 and R_3. Due to cost considerations, only 8 of the 2^7 combinations of the hard-to-change factor level combinations were used in the experiment. Completely randomizing the entire experiment was impossible because this would imply that $8 \times 3 = 24$ prototype engines would have to be built, that is one for each experimental run. However, in order to save costs, only eight prototype engines were developed, one for each of the 8 combinations of the whole plot factor levels used in the experiment. Each prototype was used under the three driving modes.

In spite of the widespread use of split-plot experiments, their design has received relatively little attention. Huang, Chen and Voelkel (1998) and Bingham and Sitter (1999) derive minimum aberration two-level fractional factorial SPDs, which are very useful for screening experiments. The use of two-level fractional factorials is, however, only possible when the number of observations available n, as well as the whole plot size k, are powers of 2. For lack of alternatives, standard response surface designs are often used to conduct split-plot experiments for fitting second order polynomials. The efficiency of several standard response surface designs under a split-plot error structure was compared by Letsinger et al. (1996). In terms of D-efficiency, the central composite and the Box-Behnken designs turn out to be the best standard designs under a split-plot error structure. However, these standard designs were developed to be used in a completely randomized experiment and not in a split-plot experiment. Therefore, it is probable that these designs are not optimal when used in a split-plot experiment. Moreover, standard response surface designs are restricted to a few sample sizes. For example, a four variable central composite design has $2^4 = 16$ factorial points, $2 \times 4 = 8$ star points and a couple of center runs. As a result, it cannot be used when the number of observations is not around 26. In addition, standard response surface designs are not flexible, for example in cases where the design region is restricted or in cases where qualitative variables are present. The algorithm of Trinca and Gilmour (2001) mentioned in the previous chapter can also be used to design split-plot experiments. Their algorithm, however, cannot be used for the present design problem because, in this chapter, the number of whole plots as well as the whole plot sizes are not dictated by the experimental situation. Instead, they can be chosen at liberty. Goos and Vandebroek (2001c) developed a design construction algorithm that computes the optimal number of whole plots and the whole plot sizes with respect to the $\mathcal{D}$-optimality criterion. Their algorithm and the computational results will be the focus of this chapter.

In the next section, we briefly describe the response surface model corresponding to a split-plot experiment. Next, the design construction algorithm of Goos and Vandebroek (2001c) is discussed and computational results are presented. The results indicate that substantial efficiency gains can be obtained by taking into account the variance-covariance structure of the experiment when designing it and that it may be more efficient to conduct a split-plot experiment than a completely randomized experiment. An example is used to demonstrate and to compare the alternative design options.

7.2   Model

The set of m experimental variables in a split-plot experiment is partitioned in two subsets. The m_w whole plot variables will be denoted by $\mathbf{w}$. The remaining $m_s = m - m_w$ variables are the sub-plot variables $\mathbf{s}$. In the chemical experiment described by Letsinger et al. (1996), there are two whole plot factors, namely the factors temperature 1 and pressure. In the prototype experiment, the factors A to G are the whole plot factors.

In general, the SPD has two types of experimental units, and therefore, it also has two randomization procedures. Firstly, each of the whole plot factor level combinations of $\mathbf{w}_i$ is assigned randomly to a whole plot, thereby generating the whole plot error variance. The second randomization consists of assigning the combinations of $\mathbf{s}_{ij}$ to the sub-plots, generating the sub-plot error variance. For a polynomial model, the jth observation within the ith whole plot can be written as

$$y_{ij} = \mathbf{f}'(\mathbf{w}_i, \mathbf{s}_{ij})\boldsymbol{\beta} + \gamma_i + \varepsilon_{ij}, \qquad (7.1)$$

where $\mathbf{f}(\mathbf{w}_i, \mathbf{s}_{ij})$ represents the polynomial expansion of the experimental variables, the $p \times 1$ vector $\boldsymbol{\beta}$ contains the p model parameters, γ_i is the random effect of the ith whole plot or the ith whole plot error, and ε_{ij} is the sub-plot error. In matrix notation, the model corresponding to a SPD is written as

$$\mathbf{Y} = \mathbf{X}\beta + \mathbf{Z}\gamma + \varepsilon, \tag{7.2}$$

where $\mathbf{X}$ represents the design matrix containing the settings of both the whole plot variables $\mathbf{w}$ and the sub-plot variables $\mathbf{s}$. The matrix $\mathbf{Z}$ is an $n \times b$ matrix of zeroes and ones assigning the n observations to the b whole plots. The random effects of the b whole plots are contained within the b-dimensional vector γ, and the random errors are contained within the n-dimensional vector ε. As in the previous chapter, it is assumed that

$$\mathrm{E}(\varepsilon) = \mathbf{0}_n \text{ and } \mathrm{Cov}(\varepsilon) = \sigma_\varepsilon^2 \mathbf{I}_n, \tag{7.3}$$

$$\mathrm{E}(\gamma) = \mathbf{0}_b \text{ and } \mathrm{Cov}(\gamma) = \sigma_\gamma^2 \mathbf{I}_b, \tag{7.4}$$

$$\text{and } \mathrm{Cov}(\gamma, \varepsilon) = \mathbf{0}_{b \times n}. \tag{7.5}$$

The ratio of the whole plot error σ_γ^2 to the sub-plot error σ_ε^2 is again denoted by η, which serves as a measure for the extent to which observations within one whole plot are correlated. We will refer to η as the degree of correlation. Finally, we will denote the size of the ith whole plot by k_i.

The analysis of a split-plot experiment is described in Section 3.3. A few specific issues concerning the analysis of a split-plot experiment receive attention in the Sections 3.5 and 6.3. The $\mathcal{D}$-optimal SPD maximizes the determinant of the information matrix

$$\mathbf{M} = \mathbf{X}'\mathbf{V}^{-1}\mathbf{X}.$$

As shown in Section 3.4, this matrix can be rewritten as a sum of outer products of vectors. This allows a fast update of the information matrix, its inverse and its determinant after a design change.

7.3   Design construction algorithm

Whereas little work has been done on the optimal design of split-plot experiments, a vast literature on the construction of discrete, tailor-made $\mathcal{D}$-optimal CRDs can be found. As could be read in Section 1.9, the most famous construction algorithms for response surface designs may be classified as exchange algorithms and include the algorithm of Fedorov (1972), the DETMAX algorithm of Mitchell (1974a) and the BLKL algorithm of Atkinson and Donev (1989). The purpose of the algorithm of Goos and Vandebroek (2001c) discussed here is to construct optimal SPDs. The structure of the algorithm is analogous to the CRD construction algorithms, but the computational work is more prohibitive due to the compound symmetric error structure. In order to evaluate the effect of design changes on the

$\mathcal{D}$-efficiency criterion, the algorithm intensively uses the update formulae derived in Section 3.4.

The algorithm starts with the generation of a non-singular n-point starting design. The design points are chosen from a predefined grid of candidate points which cover the entire design region. Part of the starting design is composed in a random fashion. Next, it is completed by sequentially adding the candidate point with the largest prediction variance. Once the starting design is complete, it is verified that it is not singular before the algorithm proceeds to the next step: the improvement of the starting design. In order to improve the starting design, the algorithm investigates whether replacing a design point by one of the candidate points leads to a larger $\mathcal{D}$-criterion value. The best possible exchange is carried out and the procedure is repeated until no further improvement can be found.

A more detailed outline of the SPD construction algorithm is displayed in Appendix A. It differs from the algorithm developed in Chapter 6 in two key aspects. Firstly, the number of whole plots b, and thus of whole plot factor levels, is not fixed in advance. Instead, it is determined optimally by the algorithm. Similarly, the whole plot sizes k_i $(i = 1, 2, \ldots, b)$ are an output of the algorithm. Secondly, only one strategy, namely the exchange of design points by candidate points, is used to improve the starting design. This is because the exchange strategy is able to find the optimal whole plot factor levels, such that the strategy of replacing whole plot factor levels is no longer useful here. The algorithm differs from a construction algorithm for completely randomized experiments in two other aspects. Firstly, the variance-covariance matrix $\mathbf{V}$ is taken into account explicitly. Secondly, the algorithm requires the specification of the unknown degree of correlation η (see also Section 6.6). Fortunately, it turns out that a reasonable guess of η is satisfactory. This will be demonstrated by means of some computational results below.

Apart from η, the input to the algorithm consists of the number of observations n, the number of tries t, the order of the model, the number of model parameters p, the number of explanatory variables m and the structure of their polynomial expansion. Finally, the whole plot and sub-plot variables need to be identified. By default, the algorithm computes the grid of candidate points $G = \{1, 2, \ldots, g\}$ as in Atkinson and Donev (1992): the design region is assumed to be hypercubic and is taken as $[-1, +1]^m$. The grid points are chosen from the 2^m, 3^m, 4^m, ... factorial design depending on whether the model contains linear, quadratic, cubic or higher order terms. Alternatively, the user can specify G if another set of candidate points is desired. This is important when the design region is hyperspherical or when it is restricted. In order to find efficient designs, the set of candidate points should certainly include corner points and cover the entire design region.

In the next section, we use an example to demonstrate the alternative design options in a specific experimental situation and to compare them in terms of $\mathcal{D}$-efficiency.

7.4   Computational results

Using the exchange algorithm, we have constructed $\mathcal{D}$-optimal SPDs for models with different numbers of variables and different numbers of whole plot and sub-plot variables under various degrees of correlation. As an illustration, we will display the 27-point three-level SPDs for a variant of the printing ink study (see Box and Draper (1987)). They will be compared to two commonly used experimental settings, both of which are based on $\mathcal{D}$-optimal CRDs. If all design points of the CRD are randomized and the levels of all experimental variables are independently reset for each run, we obtain a properly conducted CRD or PCRD. On the contrary, if all experimental runs with the same whole plot level are executed successively, that is if restricted randomization is used, we obtain an improperly conducted CRD or ICRD. In a PCRD, all observations are statistically independent. In an ICRD, observations within the same whole plot are correlated. The CRDs were computed by the algorithm of Atkinson and Donev (1989). It is also examined to what extent optimal designs are sensitive to misspecification of the degree of correlation η. Finally, we compare the efficiency of the central composite and Box-Behnken designs to that of $\mathcal{D}$-optimal SPDs.

7.4.1   Printing Ink Study

Consider an experiment in which the effect of three variables on a printing machine's ability to apply coloring inks on package labels is examined. The example comes from an exercise in Box and Draper (1987). The three factors under investigation are speed, pressure and distance. The experiment consisted of three replicates of the 3^3 factorial and was conducted using complete randomization.

Suppose that new printing machinery is installed and that a new experiment has to be carried out in order to reinvestigate the effects of the three factors. The experiment should be conducted in an economical fashion. For this reason, an SPD is to be used. In the first stage of planning the experiment, it is decided what factors will serve as whole plot and sub-plot factors respectively. This decision depends on the cost and difficulty to change the levels of the factors under investigation and should be taken in consultation with experienced production engineers. For instance, if changing the machine speed entails a large setup time and considerable costs in comparison with the factors pressure and distance, then it should be used as the only

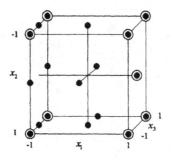

Figure 7.1: $\mathcal{D}$-optimal 27-point CRD for the full quadratic model in three variables. • is a design point, ⊚ is a design point replicated twice.

whole plot variable of the experiment. Pressure and distance would then be the sub-plot variables. If changing the machine speed and the pressure level is difficult and/or costly, then the experiment can have two whole plot variables and only one sub-plot variable. In the next section, we have developed 27-point $\mathcal{D}$-optimal SPDs for experiments with both one whole plot variable and two whole plot variables. What design should be used depends on practical and economical considerations, and on the degree of correlation η. The model of interest is the full quadratic model in three variables and the experimental region is assumed to be cubic. The design points were chosen from the 3^3 factorial design.

7.4.2   Features of $\mathcal{D}$-optimal split-plot designs

The $\mathcal{D}$-optimal 27-point CRD for the full quadratic model in three variables is displayed in Figure 7.1, in which the three variables of interest are denoted by x_1, x_2 and x_3. If a split-plot experiment with one whole plot variable is preferred, the optimal design is found in Figure 7.2. SPD1, SPD2, SPD3 and SPD4 are $\mathcal{D}$-optimal for $\eta \leq 0.3959$, $0.3959 \leq \eta \leq 0.4727$, $0.4727 \leq \eta \leq 5.7306$, and $\eta \geq 5.7306$ respectively. In the geometric representations, the whole plot variable is denoted by w, while the two sub-plot variables are denoted by s_1 and s_2. It turns out that, for $\eta \leq 0.3959$, the CRD from Figure 7.1 is optimal on the condition that the whole plot variable is assigned to the horizontal axis. For larger η, the design matrices of the CRD and the $\mathcal{D}$-optimal SPDs are different. In Figure 7.2, we see that the number of observations at $w = 0$ decreases as the degree of correlation increases and that the computed SPDs are not crossed. Computational results for full quadratic models with one whole plot variable indicate that these conclusions remain valid if the number of sub-plot variables differs from two.

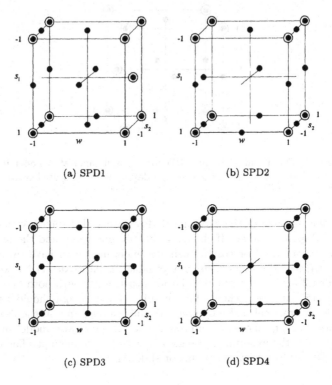

(a) SPD1   (b) SPD2

(c) SPD3   (d) SPD4

Figure 7.2: $\mathcal{D}$-optimal 27-point designs for a full quadratic model in one whole plot variable and two sub-plot variables. • is a design point, ⊙ is a design point replicated twice.

If two whole plot variables are used instead of one, the $\mathcal{D}$-optimal designs look totally different. 27-point $\mathcal{D}$-optimal SPDs for small, moderate and large η for this design problem are shown in Figure 7.3. The two whole plot variables and the sub-plot variable are denoted by w_1, w_2 and s respectively. Compared with the CRD in Figure 7.1, the SPDs have less observations in the corner points of the design region and they all have an observation in the center point. The SPDs are non-crossed and only differ in the location of one or two design points. However, no clear pattern in the designs can be observed.

For first order models with or without interactions, the design matrices of $\mathcal{D}$-optimal SPDs and CRDs are equal. The only difference between both is that, in a split-plot experiment, the assignment of the variables to the columns of the design matrix matters. For second order models, the design matrices of $\mathcal{D}$-optimal SPDs and CRDs are unequal.

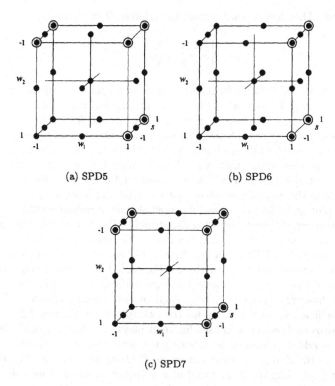

(a) SPD5   (b) SPD6

(c) SPD7

Figure 7.3: $\mathcal{D}$-optimal 27-point designs for a full quadratic model in two whole plot variables and one sub-plot variable. • is a design point, ⊙ is a design point replicated twice.

This section has illustrated that $\mathcal{D}$-optimal SPDs usually differ from the $\mathcal{D}$-optimal CRD for second order models. Both are equal for first order models with and without interactions, as well as for second order models in one whole plot variable when the degree of correlation is small. Nonetheless, even then the CRDs should be used with care since the assignment of variables matters. In the next section, it is computed to what extent $\mathcal{D}$-efficiency of experiments can be improved by taking into account the split-plot error structure in the design construction stage. It is also illustrated that the optimal SPDs are robust to misspecification of η.

7.4.3 $\mathcal{D}$-efficiency

In this section, we will compare the $\mathcal{D}$-efficiency of the three-level designs from Figures 7.1, 7.2 and 7.3 by means of the determinant of their information matrix. The $\mathcal{D}$-optimal SPD maximizes $|\mathbf{X}'\mathbf{V}^{-1}\mathbf{X}|$ for a given design problem. Therefore, we will report the relative $\mathcal{D}$-efficiencies

$$\left\{ \frac{|\mathbf{X}'\mathbf{V}^{-1}\mathbf{X}|}{|\mathbf{A}'\mathbf{\Lambda}^{-1}\mathbf{A}|} \right\}^{1/p}, \tag{7.6}$$

where $\mathbf{X}'\mathbf{V}^{-1}\mathbf{X}$ is the information matrix of a split-plot experiment with design matrix $\mathbf{X}$, and $\mathbf{A}'\mathbf{\Lambda}^{-1}\mathbf{A}$ is the information matrix of improperly conducting the CRD from Figure 7.1. The matrix $\mathbf{A}$ is thus the design matrix of the best possible 27-point completely randomized experiment for the estimation of a full quadratic model in three variables. The matrix $\mathbf{\Lambda}$ is the variance-covariance matrix resulting from an improper use of this CRD. The $\mathcal{D}$-criterion value $|\mathbf{A}'\mathbf{\Lambda}^{-1}\mathbf{A}|$ was computed using x_2 as the whole plot variable in the case of one whole plot variable, and using x_1 and x_2 as the whole plot variables in the case of two whole plot variables. In this way, the improvement, generated by using $\mathcal{D}$-optimal SPDs instead of the $\mathcal{D}$-optimal CRD in a split-plot experiment, can readily be displayed. We have used the $\mathcal{D}$-efficiency of an ICRD as a benchmark, because improperly conducting a CRD is common practice in industry. For each design considered, we have computed the $\mathcal{D}$-criterion values for degrees of correlation η between 0 and 10, holding the variance $\sigma_\epsilon^2 + \sigma_\gamma^2$ equal to one. Figures 7.4 and 7.5 show the $\mathcal{D}$-efficiencies of the SPDs for the examples from Figures 7.2 and 7.3. The horizontal reference in each figure displays the efficiency of the ICRD, which would be obtained by ignoring the structure of a split-plot experiment in the design construction stage. Both figures contain the relative efficiency of using the 3^3 factorial in a split-plot experiment as well.

The relative $\mathcal{D}$-efficiencies of the 27-point SPDs for a full quadratic model in one whole plot variable and two sub-plot variables are displayed in Figure 7.4. While both designs are equivalent at $\eta = 0$, SPD1 is more efficient than ICRD for any strictly positive η. SPD2, SPD3 and SPD4 are inferior to both ICRD and SPD1 for small η, but become substantially better as the degree of correlation increases. Except for η close to zero, efficiency gains lie between 1% and 2% and are robust against misspecification of the degree of correlation. The robustness comes from the fact that the SPDs are optimal for a range of degrees of correlation. Slightly misspecifying η therefore results in the same design. The figure also demonstrates that efficient designs for large η do not perform well when $\eta = 0$, that is when used in a completely randomized experiment. The 3^3 factorial turns out to be a poor alternative to the SPDs and even to ICRD.

Figure 7.5 shows a similar picture for the SPDs with two whole plot variables. At $\eta = 0$, ICRD is $\mathcal{D}$-optimal, but for strictly positive η, it is overtaken by the SPDs. For η not too close to zero, efficiency gains lie between 1% and 5%. Like in the one whole plot variable case, the SPDs are optimal for a range of degrees of correlation. Again, it turns out that SPDs with an

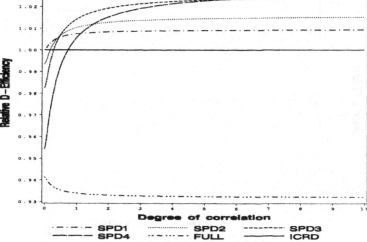

Figure 7.4: Relative $\mathcal{D}$-efficiencies of the 27-point CRD and SPDs from Figures 7.1 and 7.2, and the 3^3 factorial for the full quadratic model in one whole plot variable and two sub-plot variables for different degrees of correlation.

excellent relative efficiency for large η are inefficient when η is small. The 3^3 factorial remains a bad alternative to the SPDs, but has become better than ICRD at larger values of η.

7.4.4 Comparison to a completely randomized experiment

In Figure 7.6, the $\mathcal{D}$-efficiency of properly conducting a CRD (PCRD) is compared to using a split-plot experiment with one and with two whole plots. In other words, we investigate whether it is statistically efficient to do the effort to perform a completely randomized experiment and to obtain independent observations or whether it is better to perform a split-plot experiment and to obtain correlated observations. For this purpose, we have computed the relative $\mathcal{D}$-efficiencies

$$\left\{ \frac{|\mathbf{X'V^{-1}X}|}{|\mathbf{A'A}|} \right\}^{1/p}. \tag{7.7}$$

of the best SPDs with one and two whole plots (SPDW1 and SPDW2 respectively) and of improperly conducting a CRD with one and with two whole plot variables (ICRDW1 and ICRDW2 respectively). The relative efficiencies were computed for η-values between 0 and 10, holding $\sigma_\varepsilon^2 + \sigma_\gamma^2 = 1$. The denominator in (7.7) is the efficiency of properly conducting the CRD from Figure 7.1. The numerator is the efficiency of the split-plot experiment

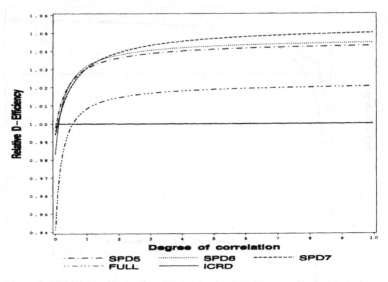

Figure 7.5: Relative $\mathcal{D}$-efficiencies of the 27-point CRD and SPDs from Figures 7.1 and 7.3, and the 3^3 factorial for the full quadratic model in two whole plot variables and one sub-plot variable for different degrees of correlation.

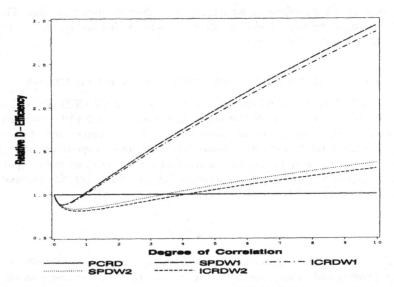

Figure 7.6: Comparison of the $\mathcal{D}$-efficiency of the SPDs from Figures 7.2 and 7.3 and proper and improper use of the CRD from Figure 7.1 for the full quadratic model in three variables.

under investigation. For small degrees of correlation, conducting a PCRD is statistically more efficient than a split-plot experiment. However, as η rises, the opposite is true. When η exceeds unity, conducting an SPD with one whole plot variable becomes much more efficient than PCRD. When η grows larger than 3, the same goes for SPDs with two whole plot variables. The efficiency of the ICRDs closely follows that of the corresponding SPDs. Except for small degrees of correlation, conducting an SPD with one whole plot variable is clearly more efficient than using two whole plot variables. These results illustrate that inducing correlation between observations may have a beneficial effect on $\mathcal{D}$-optimality. For saturated designs, it can even be shown that correlated observations always lead to better $\mathcal{D}$-criterion values. A proof is given in Appendix B.

The computational results have shown that efficiency gains over existing approaches can be achieved by taking into account the split-plot format in which an experiment will be performed. Moreover, Figure 7.6 has illustrated that performing a split-plot experiment may be more efficient than conducting a completely randomized experiment. This result implies that split-plot experiments are not only easier to perform, but in many cases also statistically more efficient.

7.4.5   Comparison to standard response surface designs

For some specific sample sizes n, it is meaningful to compare $\mathcal{D}$-optimal SPDs with standard response surface designs. As Letsinger et al. (1996) pointed out, the central composite designs (CCDs) and the Box-Behnken designs provide efficient design options for split-plot experiments. Consider a full quadratic model in four variables. Apart from the center runs, the four variable CCD and Box-Behnken design both have 24 design points. For both a hypercubic and a hyperspherical design region, we will compare the 25-point $\mathcal{D}$-optimal SPDs to these two standard designs augmented by one center run.

Let us first consider a hypercubic design region and assume that the axial portion of the CCD is determined by $\alpha = 1$. When there is one whole plot variable, the determinant of the CCD with one center run is three to four times smaller than that of the 25-point $\mathcal{D}$-optimal SPD for $0 \leq \eta \leq 10$. When there are two whole plot variables, the $\mathcal{D}$-criterion value of the CCD with one center run is even three to six times smaller than that of the 25-point $\mathcal{D}$-optimal SPD. The Box-Behnken design is inferior to the CCD and the SPDs.

Now consider a hyperspherical design region with radius two. Assume that the factorial portion of the CCD has levels ± 1 and that the axial portion is determined by $\alpha = 2$. Apart from the center point, all design points of the

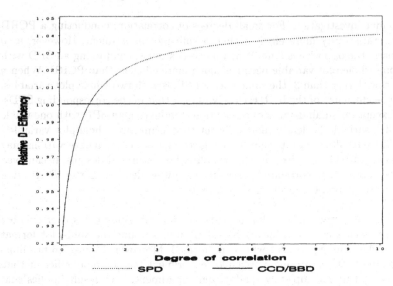

Figure 7.7: Relative $\mathcal{D}$-efficiencies of the CCD and the BBD with one center run for a full quadratic model in two whole plot and two sub-plot variables.

CCD then lie on the surface of the experimental region. In order to obtain the same effect with the Box-Behnken design, we multiply its factor levels by $\sqrt{2}$. Letsinger et al. (1996) discovered that the CCD is superior to the Box-Behnken design when there is one whole plot variable and that both are close competitors when there are two whole plot variables. Let us now compare both designs to the $\mathcal{D}$-optimal SPDs. For the computation of the SPDs, we have used the union of the points of the CCD, the Box-Behnken design and the 3^4 factorial as the set of possible design points. This gave us 113 candidate points and seven levels for each of the experimental variables, namely 0, ±1, $\pm\sqrt{2}$ and ±2. When there is one whole plot variable, the CCD with one center run is the $\mathcal{D}$-optimal design. When there are two whole plot variables, the CCD is optimal for small degrees of correlation. However, as the degree of correlation approaches unity, the CCD is no longer the best design option. It turns out that the Box-Behnken design is almost as efficient as the CCD when there are two whole plot variables, but that it is inferior when there is only one whole plot variable. This is illustrated in Figure 7.7.

These results indicate that, for a spherical design region, the CCD with one center run is often a good design option for a split-plot experiment. However, this is no longer true when the number of center points is increased, for example to obtain a uniform precision (7 center runs) or an orthogonal (12 center runs) rotatable CCD. The negative impact of increasing the number of center runs on the $\mathcal{D}$-efficiency of a completely randomized CCD is described by Lucas (1974, 1976, 1977). Similarly, we have found that using a CCD with a large number of center runs in a split-plot experiment is not the best design option in terms of $\mathcal{D}$-efficiency.

Appendix A. The construction algorithm

We denote the set of g candidate points and the set of design points by G and D respectively. We denote the total number of observations by n and the number of observations in each grid point i by n_i. Note that $i \in D$ if and only if $n_i > 0$. For simplicity, we will denote the information matrix $\mathbf{X'V^{-1}X}$ by $\mathbf{M}$. The singularity when $n < p$ is overcome by using $\mathbf{M} + \omega\mathbf{I}$ instead of $\mathbf{M}$ with ω a small positive number. We refer to the $\mathcal{D}$-criterion value of a design D as $\mathcal{D}$. The number of observations in point i in the current optimal design D^* is denoted by n_i^* and the corresponding criterion value is denoted by $\mathcal{D}^*$. Finally, we denote the number of tries by t and the number of the current try by t_c. The algorithm starts by specifying the set of grid points $G = \{1, 2, \ldots, g\}$ and proceeds as follows:

1. Set $\mathcal{D}^* = 0$ and $t_c = 1$.

2. Set $D = \emptyset$, $\mathbf{M} = \omega\mathbf{I}$ and $\forall i \in G : n_i = 0$.

3. Generate starting design.
    (a) Randomly choose j.
    (b) Do j times:
        i. Randomly choose $i \in G$.
        ii. Set $n_i = n_i + 1$ and $D = D \cup \{i\}$.
        iii. Update $\mathbf{M}$.
    (c) Do $n - j$ times:
        i. Determine $i \in G$ with largest prediction variance.
        ii. Set $n_i = n_i + 1$ and $D = D \cup \{i\}$.
        iii. Update $\mathbf{M}$.

4. Compute $\mathbf{M}$ and $\mathcal{D}$. If $\mathcal{D} = 0$, then go to step 2, else continue.

5. Evaluate exchanges.
    (a) Set $\delta = 1$.
    (b) $\forall i \in G, \forall j \in D, i \neq j :$
        i. Compute the effect $\delta_j^i = \mathcal{D'}/\mathcal{D}$ of exchanging j by i.
        ii. If $\delta_j^i > \delta$, then $\delta = \delta_j^i$ and store i and j.

6. If $\delta > 1$, then go to step 7, else go to step 8.

7. Exchange j by i. Set $n_i = n_i + 1$ and $n_j = n_j - 1$. Update $\mathbf{M}$, D and $\mathcal{D}$. Go to step 5.

8. If $\mathcal{D} \geq \mathcal{D}^*$, then $\mathcal{D}^* = \mathcal{D}$, $D^* = D$, $\forall i \in G: n_i^* = n_i$.

9. If $t_c < t$, then $t_c = t_c + 1$ and go to step 2, else stop.

Appendix B. Saturated designs with correlated observations

Let $\mathbf{X}$ denote the design matrix of a given experiment. If a completely randomized experiment is conducted, the $\mathcal{D}$-criterion value is given by $|\mathbf{X}'\mathbf{X}|$. If the experiment is conducted as as split-plot experiment, the observations within each whole plot are correlated. Let $\mathbf{R}$ denote the correlation matrix of the observations. The $\mathcal{D}$-criterion value of this experiment is proportional to

$$
\begin{aligned}
|\mathbf{X}'\mathbf{R}^{-1}\mathbf{X}| &= |\mathbf{X}'|\,|\mathbf{R}^{-1}|\,|\mathbf{X}|, \\
&= |\mathbf{R}^{-1}|\,|\mathbf{X}'|\,|\mathbf{X}|, \\
&= |\mathbf{R}|^{-1}|\mathbf{X}'\mathbf{X}|.
\end{aligned}
$$

Since $|\mathbf{R}| \leq 1$ and, consequently, $|\mathbf{R}^{-1}| \geq 1$, the $\mathcal{D}$-criterion value for the split-plot experiment is larger than or equal to the value for the completely randomized experiment. Equality holds only if the correlation matrix $\mathbf{R}$ is diagonal. It is clear that the above proof holds for any correlation matrix.

8
Optimal Split-Plot Designs

Split-plot designs are heavily used in industry, especially when factor levels are difficult or costly to change or to control. This is because this type of design avoids too many changes of the whole plot factor levels, which leads to considerable savings in cost and time. The purpose of this chapter on split-plot designs is twofold. Firstly, we investigate to what extent the designs of Chapter 7 can be improved if the number of whole plots is increased. Secondly, we compare split-plot designs to completely randomized experiments in terms of $\mathcal{D}$-efficiency. It turns out that the former are often more efficient than the latter.

8.1 Introduction

As already indicated in the previous chapters, split-plot designs (SPDs) have become increasingly popular in industry especially when it is difficult or costly to control or to change the levels of some of the experimental factors. Typical examples of such factors are pressure, humidity and process temperature. Rather than conducting a completely randomized experiment in which pressure has to be moved back and forth according to the randomization scheme, successively executing experimental runs with equal pressure is preferred by the experimenter. Alternatively, experimenters often group all experimental runs with the same temperature level and execute them simultaneously in one furnace. In doing so, the number of changes in the levels of the hard-to-change factors is limited to the number of levels. As a result, by using an SPD in the presence of hard-to-change factors, the ease of experimentation is significantly increased and precious time and resources can be saved. However, this should not be the only reason to consider using a split-plot experiment.

In this chapter, we will show that split-plot designs are often statistically much more efficient than completely randomized experiments. Contrary to the split-plot designs described in Chapter 7, the SPDs described here allow more factor level changes. The benefits of this approach are fivefold. Firstly, the statistical efficiency of the experiment is increased. Next, an increased number of whole plots ensures an improved control of variability and provides a better protection against possible trend effects. Increasing the number of level changes also protects the experimenter against systematic errors that may occur when something goes wrong at a certain hard-to-change factor level. In addition, more degrees of freedom are available for the estimation of the whole plot error. Finally, the number of factor level changes is mostly smaller than in a completely randomized design (CRD). For all these reasons, the SPDs discussed here should not only be considered when some of the experimental factors are hard-to-change, but also when no hard-to-change factors are present.

The SPDs derived in this chapter differ from those computed in Chapter 6 in that the number of whole plots and the whole plot size are not fixed in advance, but they are determined by the design construction algorithm. As a result, the set of split-plots considered here is, in a way, a generalization of both the constrained split-plot designs from Chapter 6 and the SPDs derived in Chapter 7.

It is clear that a completely randomized experiment can be seen as a split-plot experiment in which each run is assigned to a different whole plot. Each whole plot then contains only one observation so that $\mathbf{V}$ is a diagonal matrix and the sub-plot error variance σ_ε^2 and the whole plot error variance σ_γ^2 cannot be distinguished from each other.

In the next section, the approach advocated in this chapter is motivated and illustrated by means of an example. In Section 8.3, an efficient algorithm for the construction of $\mathcal{D}$-optimal SPDs is presented. Finally, the computational results and the practical consequences are discussed in Section 8.4.

Table 8.1: Steel experiment with four independent heatings.

Whole plot 1			Whole plot 3		
Temp	Or	Alloy	Temp	Or	Alloy
675	1	1	725	1	1
675	1	2	725	1	2
675	1	3	725	1	3
675	2	1	725	2	1
675	2	2	725	2	2
675	2	3	725	2	3
Whole plot 2			Whole plot 4		
Temp	Or	Alloy	Temp	Or	Alloy
700	1	1	750	1	1
700	1	2	750	1	2
700	1	3	750	1	3
700	2	1	750	2	1
700	2	2	750	2	2
700	2	3	750	2	3

8.2 Increasing the number of level changes

In this section, we extend the class of SPDs considered in the previous chapter by allowing more changes in the whole plot factor levels. Firstly, we illustrate the modification by means of an example. Next, we motivate the usefulness of the new approach and the importance of the corresponding design problem.

8.2.1 Example

In order to clarify the purpose of this chapter, consider a factorial experiment from the steel industry described by Andersen and McLean (1974). The experiment was meant to investigate the impact of furnace temperature and orientation within the furnace on the strength of three alloys. We assume the model of interest contains all linear factor effects and the quadratic effect of the factor temperature. Two different orientations (random and aligned) and four different temperatures (675, 700, 725 and 750 degrees) were used. For the sake of experimental ease, all $2 \times 3 = 6$ experimental runs for which the temperature was at the same level were conducted simultaneously in the furnace. In doing so, the furnace is heated only four times and the experiment involves four whole plots. It is obvious that there is a one-to-one relation between the whole plot factor levels (temperatures) and the whole plots (heatings). The design of the split-plot experiment is displayed in Table 8.1.

In this chapter, we allow the whole plot factor temperature to be reset more often. Thereby, we increase the number of whole plots in the experiment. Each whole plot then contains less observations for a given number of ex-

Table 8.2: Steel experiment with six independent heatings.

Whole plot 1			Whole plot 3			Whole plot 5		
Temp	Or	Alloy	Temp	Or	Alloy	Temp	Or	Alloy
675	1	1	700	1	3	750	1	2
675	2	1	700	2	2	750	2	2
675	1	2	700	1	1	750	1	3
675	2	3	700	2	2	750	2	1

Whole plot 2			Whole plot 4			Whole plot 6		
Temp	Or	Alloy	Temp	Or	Alloy	Temp	Or	Alloy
675	1	3	725	1	1	750	1	1
675	2	2	725	2	3	750	2	3
675	1	3	725	1	2	750	1	2
675	2	1	725	2	3	750	2	1

perimental runs. As before, exactly one temperature corresponds to each whole plot (heating). However, more than one whole plot corresponds to each whole plot factor level (temperature). For instance, an alternative to the original experiment is to heat the furnace eight times and to conduct three observations simultaneously at each heating. Although this experimental setup is more cumbersome than the original, it is still much easier to conduct than a completely randomized experiment because the latter would require $3 \times 4 \times 2 = 24$ instead of 8 independent heatings. Moreover, increasing the number of whole plots yields designs that are statistically more efficient than the original experiment. As an illustration, a $\mathcal{D}$-optimal SPD with six whole plots is shown in Table 8.2. For slightly correlated responses ($\eta = 0.1$), the $\mathcal{D}$-criterion value is improved by 8.15%. For larger degrees of correlation, the improvement is even more pronounced and the resulting experiment is even slightly better than a completely randomized experiment with the design points of the original experiment. In the sequel of the chapter, we will show that situations in which completely randomized experiments are outperformed by SPDs in terms of $\mathcal{D}$-efficiency are numerous when the number of whole plots is slightly increased. We will now further motivate the use of increasing the number of changes in the whole plot factor levels.

8.2.2 Motivation

Goos and Vandebroek (2001c) found that $\mathcal{D}$-optimal SPDs can be more efficient than $\mathcal{D}$-optimal completely randomized experiments (see Section 7.4.4). Moreover, they proved that design efficiency always benefits from correlated observations when the designs are saturated (see Appendix B of Chapter 7). Similarly, Anbari and Lucas (1994) and Lucas and Ju (2002) show that properly arranging the runs of a two-level factorial design in whole plots is a better approach in terms of $\mathcal{G}$-efficiency than running a completely randomized design and than running the design in a random order without resetting the levels of the hard-to-change factors. Lucas and Ju (2002) refer to the latter approach by the name random run order. A strong corollary of these results is that the design construction algorithms described in Section 1.9 do not generate the best design for a given experimental situation, but only the best completely randomized design. In many cases, a more efficient (split-plot) design can be found.

Using the algorithm of Goos and Vandebroek (2001c) discussed in the previous chapter, we obtained strong indications that substantial efficiency gains could be realized by relaxing the one-to-one relation between the combinations of whole plot factor levels and the whole plots. By choosing design points from a 21×21 grid instead of a 3×3 grid on $\chi = [-1,1]^2$, the algorithm produced much better designs in terms of $\mathcal{D}$-efficiency. Consider, for example, the $\mathcal{D}$-optimal SPDs for $\eta = 1$ for a full quadratic model in one whole plot variable w and one sub-plot variable s in Figure 8.1 obtained by using the default 3×3 grid and the finer 21×21 grid on χ. It turns out that the SPD in Figure 8.1b is 24% more efficient than that in Figure 8.1a. The former design possesses a larger number of whole plot levels, and thus of whole plots, than the latter. Besides -1, 0 and +1, the design construction algorithm also chooses ± 0.1, -0.2 and ± 0.9 as whole plot levels. The optimal sub-plot levels remain 0 and ± 1. This result suggests that the optimal whole plot levels are indeed at the 0 and ± 1 levels, but that not all observations at a certain whole plot level should be put in the same whole plot. Instead, it strongly recommends the use of more whole plots in order to decrease the number of correlated observations.

Another advantage of increasing the number of whole plots is that more degrees of freedom are available for the estimation of the whole plot error. The whole plot degrees of freedom are given by the difference between the rank of $[\mathbf{X} \quad \mathbf{Z}]$ and the rank of $\mathbf{X}$. All other things being equal, increasing the number of whole plots increases the number of columns in $\mathbf{Z}$ and thus the rank of $[\mathbf{X} \quad \mathbf{Z}]$. The sub-plot error degrees of freedom are given by the number of observations n minus the rank of $[\mathbf{X} \quad \mathbf{Z}]$. As a result, increasing the number of whole plots leads to a decrease in the degrees of freedom available for sub-plot error estimation. Ganju and Lucas (1999) plead for properly designed split-plot experiments that allow decent estimation of both error components. Of course, better estimates for the variance components will lead to better estimates of the model parameters too.

Box and Jones (1992), Davison (1995) and Ganju and Lucas (1997) point out that a loss of precision in the whole plot coefficients is incurred when a SPD is used. However, it turns out that a small increase in the number of whole plots largely solves this problem. This was already illustrated in Section 6.3.

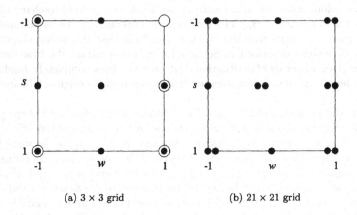

(a) 3 × 3 grid   (b) 21 × 21 grid

Figure 8.1: $\mathcal{D}$-optimal 15-point SPDs for the full quadratic model in two variables for $\eta = 1$. Candidate design points are points from a grid on $\chi = [-1, 1]^2$. • is a design point, ⊙ is a design point replicated twice and ○ is a design point replicated three times.

Finally, increasing the number of factor level changes ensures a better randomization and reduces the risk that systematic errors or trend effects distort the results of the experiment. For example, if something goes wrong at a given level of a hard-to-change factor, this will influence all the observations at this level since it is not independently reset for each experimental run. As experiments in the presence of hard-to-change factors typically use a minimum of factor levels, this may have dramatic consequences for the analysis of the experiment. In other words, even if factor levels are hard to change, it can be worthwhile to increase the number of whole plots.

8.3   Design construction

The construction of $\mathcal{D}$-optimal SPDs is more complicated than the computation of $\mathcal{D}$-optimal CRDs because design efficiency depends on the compound symmetric error structure. In addition, the fact that more than one whole plot per whole plot factor level is allowed does not simplify the construction of $\mathcal{D}$-optimal SPDs.

As in Chapter 7, the algorithm presented here autonomously computes the optimal whole plot sizes and gives the optimal allocation of sub-plot levels to each whole plot for a given number of observations n and degree of correlation η. In addition, the number of whole plots per whole plot factor level combination needs to be determined as well. The algorithm starts with the computation of a starting design. First, a number of points are randomly selected from the list of candidates. The starting design is then completed by sequentially adding the candidate point with the largest prediction variance. Each time a point was added during the starting phase, it was randomly assigned to an existing or a new whole plot. In order to improve the starting design, three alternative strategies are evaluated for each combination of a design point and a point from the candidate list. In each of the alternatives, the design point is removed from the design and the candidate point is added to the design. Firstly, the candidate point entering the design can be assigned to the same whole plot as the point removed from the design. Of course, this is only possible when both points have the same whole plot factor level. Secondly, the entering point can be assigned to another whole plot than that of the removed point, independent of its whole plot factor level. Finally, the entering design point can also be assigned to a new whole plot. When all possible exchanges have been evaluated, the best one is carried out. This process is then repeated until no further improvement can be made. A more detailed outline of the algorithm can be found in the Appendix to this chapter.

The input to the algorithm consists of the desired number of observations n, the number of tries, the order of the model, the number of model parameters p, the number of explanatory variables m and the structure of their polynomial expansion f. The whole plot and sub-plot factors need to be identified and an estimate of the degree of correlation η must be provided as well (see also Section 6.6). Since the computed designs are optimal for a range of η-values, a reasonable guess is satisfactory for the purpose of design construction.

In the next section, computational results demonstrate the benefits of using split-plot designs with a larger number of whole plots instead of split-plot designs with a small number of whole plots and instead of completely randomized designs.

8.4   Computational results

In this section, a small example illustrates the features of the optimal SPDs and the substantial gains that can be achieved by conducting split-plot experiments instead of completely randomized experiments. The effects of limiting the number of whole plots and/or imposing constraints on the number of sub-plots within each whole plot for economical or practical reasons is investigated as well. In addition, a factorial experiment shows that large efficiency gains can be realized for pure linear models, linear models with two-factor interactions and quadratic models.

Consider a full quadratic model in one whole plot variable w and one sub-plot variable s. For $0 \leq \eta \leq 0.7011$, $0.7011 \leq \eta \leq 0.9113$ and $\eta \geq 0.9113$, the 10-point $\mathcal{D}$-optimal SPDs obtained by the algorithm described in the previous section are given by

$$
\begin{bmatrix}
-1 & -1 \\
-1 & +1 \\
\hdotsfor{2} \\
-1 & 0 \\
\hdotsfor{2} \\
-1 & +1 \\
\hdotsfor{2} \\
0 & -1 \\
\hdotsfor{2} \\
0 & 0 \\
\hdotsfor{2} \\
0 & +1 \\
\hdotsfor{2} \\
+1 & 0 \\
\hdotsfor{2} \\
+1 & -1 \\
+1 & +1
\end{bmatrix},
\quad
\begin{bmatrix}
-1 & -1 \\
-1 & +1 \\
\hdotsfor{2} \\
-1 & 0 \\
-1 & +1 \\
\hdotsfor{2} \\
0 & -1 \\
\hdotsfor{2} \\
0 & 0 \\
\hdotsfor{2} \\
0 & +1 \\
\hdotsfor{2} \\
+1 & 0 \\
\hdotsfor{2} \\
+1 & -1 \\
+1 & +1
\end{bmatrix}
\quad \text{and} \quad
\begin{bmatrix}
-1 & -1 \\
-1 & +1 \\
\hdotsfor{2} \\
-1 & 0 \\
\hdotsfor{2} \\
0 & 0 \\
\hdotsfor{2} \\
0 & +1 \\
\hdotsfor{2} \\
+1 & -1 \\
+1 & +1 \\
\hdotsfor{2} \\
+1 & -1 \\
+1 & 0 \\
+1 & +1
\end{bmatrix}
\qquad (8.1)
$$

respectively. We will refer to these three designs as SPD1, SPD2 and SPD3 respectively. These designs have a couple of striking features that hold generally. Firstly, the optimal number of whole plots decreases as the correlation increases. The 10-point designs shown in (8.1) have eight, seven and six whole plots. Apparently, the higher the correlation between observations within the same whole plot, the better it is to group more experimental runs and thereby induce more correlated observations. Otherwise, the lower the correlation, the more the optimal design will resemble a completely randomized experiment. Secondly, observations at the zero levels of the whole plot variables are assigned to separate whole plots, so that whole plots containing more than one observation only occur at $\mathbf{w} = \pm 1$. Thirdly, the designs are neither crossed nor balanced.

In order to obtain the optimal SPDs in (8.1), no constraint was imposed on the number of whole plots. However, limited resources often impose such constraint. Suppose, for instance, that no more than three whole plots are allowed. In this case, the $\mathcal{D}$-optimal SPDs are given in Figure 8.2. We will refer to these alternative design options as RSPD1 and RSPD2 respectively. Another possible constraint could be that the number of whole plots should be equal to five and that the number of runs within each whole plot should be equal to two. $\mathcal{D}$-optimal designs for $\eta \leq 1.965$ and $\eta \geq 1.965$ under these

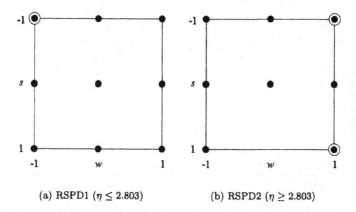

(a) RSPD1 $(\eta \leq 2.803)$ (b) RSPD2 $(\eta \geq 2.803)$

Figure 8.2: $\mathcal{D}$-optimal 10-point SPDs with three whole plots for the full quadratic model in one whole plot variable w and one sub-plot variable s. • is a design point, ⊙ is a design point replicated twice.

constraints are given by

$$
\begin{bmatrix}
-1 & -1 \\
-1 & +1 \\
\hdashline
-1 & 0 \\
-1 & +1 \\
\hdashline
0 & -1 \\
0 & 0 \\
\hdashline
+1 & 0 \\
+1 & +1 \\
\hdashline
+1 & -1 \\
+1 & +1
\end{bmatrix}
, \text{ and }
\begin{bmatrix}
-1 & -1 \\
-1 & +1 \\
\hdashline
-1 & 0 \\
-1 & +1 \\
\hdashline
0 & -1 \\
0 & 0 \\
\hdashline
+1 & -1 \\
+1 & +1 \\
\hdashline
+1 & -1 \\
+1 & +1
\end{bmatrix}
\tag{8.2}
$$

respectively. We denote these balanced split-plot designs as BSPD1 and BSPD2 respectively. BSPD1 and BSPD2 were computed using the algorithm described in Chapter 6.

In order to visualize the superiority of split-plot experiments to completely randomized experiments and to investigate the effect of imposing constraints on the design, we have computed the relative $\mathcal{D}$-efficiencies

$$
\left\{ \frac{|\mathbf{X}'\mathbf{V}^{-1}\mathbf{X}|}{|\mathbf{A}'\mathbf{A}|} \right\}^{1/p},
\tag{8.3}
$$

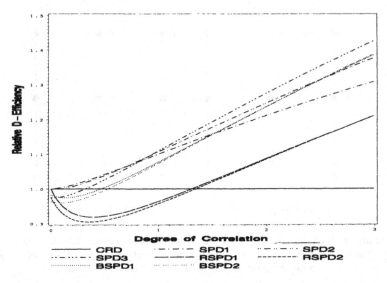

Figure 8.3: Comparison of the $\mathcal{D}$-efficiency of the 10-point SPDs in Equation (8.1), Figure 8.2 and Equation (8.2) to the $\mathcal{D}$-optimal CRD for the full quadratic model in one whole plot and one sub-plot variable for various degrees of correlation.

where $\mathbf{X}$ is the design matrix of the split-plot experiment under consideration, $\mathbf{V}$ is its variance-covariance matrix, and $\mathbf{A}$ is the design matrix of the $\mathcal{D}$-optimal CRD. The relative efficiencies were computed for η-values between 0 and 3, i.e. from zero correlation to largely correlated observations, holding $\sigma_\varepsilon^2 + \sigma_\gamma^2 = 1$. The numerator of (8.3) gives the $\mathcal{D}$-criterion value of the SPD under investigation. The denominator gives the $\mathcal{D}$-criterion value of the $\mathcal{D}$-optimal CRD when $\sigma_\varepsilon^2 = 1$. The design points of the $\mathcal{D}$-optimal CRD are the same as those of SPD1, SPD2 and RSPD1.

The relative efficiencies of SPD1, SPD2, SPD3, RSPD1, RSPD2, BSPD1 and BSPD2 are displayed in Figure 8.3. SPD1 and SPD2 are superior to the CRD for degrees of correlation between 0 and 3. SPD3 is better than the CRD when $\eta > 0.323$. The gains that can be obtained by using split-plot designs amount to 10% when $\eta = 1$ and to 43% when $\eta = 3$. For larger degrees of correlation, the gains are even more elevated. Restricting the number of whole plots b and/or the whole plot sizes k_i has a substantial negative impact on design efficiency. This is especially true for RSPD1 and RSPD2 which possess only three whole plots. Despite the fact that their whole plot sizes are constrained to two, BSPD1 and BSPD2 perform considerably better because they possess a larger number of whole plots. BSPD1 and BSPD2 outperform the CRD for η-values larger than 0.5.

η	Pure Linear			Linear + interactions			Quadratic		
	n	b	rel eff	n	b	rel eff	n	b	rel eff
0.1	5	4	1.01	8	4	1.02	12	6	1.01
0.25		4	1.02		4	1.05		6	1.05
0.5		3	1.06		4	1.11		6	1.11
0.75		3	1.11		3	1.20		5	1.19
1		3	1.15		3	1.28		5	1.28
2		3	1.34		3	1.62		5	1.61
0.1	8	4	1.00	12	6	1.02	18	11	1.02
0.25		4	1.02		5	1.05		10	1.05
0.5		4	1.06		5	1.13		8	1.12
0.75		4	1.11		4	1.22		8	1.20
1		4	1.15		4	1.31		8	1.29
2		4	1.34		4	1.67		7	1.63
0.1	10	7	1.01	16	7	1.02	24	14	1.01
0.25		7	1.03		7	1.05		12	1.04
0.5		7	1.07		7	1.12		11	1.12
0.75		6	1.12		6	1.21		10	1.20
1		6	1.17		6	1.29		10	1.29
2		6	1.35		6	1.64		10	1.62

Table 8.3: Properties of $\mathcal{D}$-optimal designs for a pure linear model, a linear model with interactions and a full quadratic model in one whole plot variable and two sub-plot variables.

The negative impact of restricting the number of whole plots can also be seen from Figure 8.4, where the relative efficiency of 10-point SPDs with at most 3, 4, 5, 6, 7 or 8 whole plots is displayed. It turns out that the effect of the restriction is largest when η is small. This is logical since it is for small η that the optimal SPDs possess the largest number of whole plots.

We have performed a factorial experiment to investigate the role of several model characteristics on the properties of the $\mathcal{D}$-optimal designs. We have computed designs for a pure linear model, for a linear model with two-factor interactions and for a full quadratic model in three variables for six different degrees of correlation ($\eta = 0.1, 0.25, 0.5, 0.75, 1, 2$). For each combination, we have computed a nearly saturated design, a design with twice as much observations and one in between. Also, we have investigated the effect of the number of whole plot variables. The results are shown in Tables 8.3 and 8.4. The first column of the tables shows the degree of correlation as measured by η. For each model under investigation, the number of design points n, the number of whole plots b and the relative $\mathcal{D}$-efficiency of the SPDs with respect to the $\mathcal{D}$-optimal CRD are displayed. The factorial experiment confirms the main results of the 10-point example in (8.1) and provides additional insights.

Table 8.3 shows that efficiency gains for a pure linear model lie between 0% and 35% and between 1% and 63% for a full quadratic model when

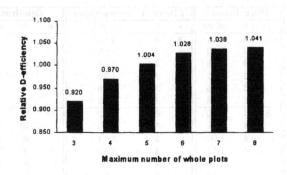

(a) $\eta = 0.5$

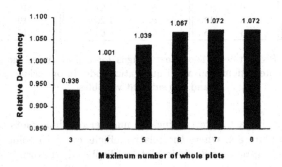

(b) $\eta = 0.75$

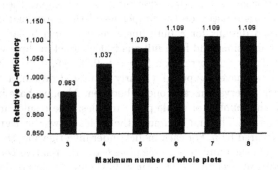

(c) $\eta = 1$

Figure 8.4: Effect of restricting the number of whole plots on the relative $\mathcal{D}$-efficiency for a full quadratic model in one whole plot and one sub-plot variable.

	Pure Linear			Linear + interactions			Quadratic		
η	n	b	rel eff	n	b	rel eff	n	b	rel eff
0.1	5	5	1.00	8	8	1.00	12	11	1.00
0.25		5	1.00		8	1.00		11	1.00
0.5		5	1.00		4	1.01		8	1.02
0.75		5	1.00		4	1.04		8	1.04
1		5	1.00		4	1.07		8	1.06
2		4	1.05		4	1.20		7	1.16
0.1	8	8	1.00	12	12	1.00	18	18	1.00
0.25		8	1.00		10	1.00		14	1.00
0.5		8	1.00		8	1.02		14	1.02
0.75		8	1.00		8	1.06		14	1.05
1		8	1.00		8	1.09		14	1.07
2		7	1.01		8	1.23		13	1.17
0.1	10	10	1.00	16	16	1.00	24	24	1.00
0.25		10	1.00		13	1.00		20	1.00
0.5		10	1.00		12	1.02		20	1.02
0.75		10	1.00		12	1.04		18	1.04
1		9	1.00		11	1.07		18	1.06
2		8	1.04		9	1.20		15	1.15

Table 8.4: Properties of $\mathcal{D}$-optimal designs for a pure linear model, a linear model with interactions and a full quadratic model in two whole plot variables and one sub-plot variable.

there is one whole plot variable. It turns out that the higher the correlation, the larger the efficiency gain of using an SPD instead of a CRD. Also, the efficiency gains for a linear model with interaction effects and for a full quadratic model are substantially higher than for a pure linear model. For all models, the number of whole plots in the optimal designs decreases as the correlation increases. In other words, the higher the degree of correlation, the more observations will tend to be grouped in the same whole plot. Table 8.4 shows that efficiency gains in the presence of two whole plot factors are smaller. This is because the larger number of whole plot factor levels reduces the possibility to group observations and to benefit from the correlation. It turns out that the CRD is the optimal design option in many cases when there are two whole plot variables. In Table 8.4, these are the cases in which the number of whole plots b equals the number of observations n. However, the number of whole plots in the optimal design becomes smaller than the number of observations as the model moves from pure linear to quadratic and as the degree of correlation increases. The results also indicate that observations at the zero level of the whole plot variables are seldom grouped in one whole plot. In holding these observations statistically independent, the variance of the parameters corresponding to the quadratic whole plot factor effects is kept low. In general, the optimal designs are not crossed. For the pure linear model and for the linear model with two-way interactions the optimal SPDs may be balanced, especially if n is a power of 2. The 8-point $\mathcal{D}$-optimal SPD for the pure linear model in one whole plot variable and two sub-plot variables is given by the 2^3 factorial design or by two replicates of a 2^{3-1} fractional factorial design. The optimal 8-point SPD for a linear model with interactions has the points of the 2^3 full factorial experiment as its design points. Whereas the eight observations for the pure linear model are allocated to four whole plots of two observations each no matter what value for η is specified, the allocation of observations to the whole plots depends on the degree of correlation and makes a substantial difference in design efficiency when interactions are included in the model. For small degrees of correlation, the eight observations are divided in four whole plots of two observations. However, for $\eta > 0.5$ only three whole plots are used: one with four observations and two with two observations. Finally, all other things being equal, the efficiency gains do not depend on the number of observations available.

8.5   Discussion

Split-plot designs are heavily used in industry, especially when hard-to-change factors are present. A key property of split-plot designs is the fact that the levels of the hard-to-change factors —also referred to as whole plot factors— are changed as little as possible. In doing so, the ease of experimenting is increased and time and costs are saved.

In this chapter, it is shown that split-plot designs may be a much better design option than completely randomized designs. However, this will only be the case when the observations in the split-plot experiment are truly correlated. In addition, the most striking efficiency gains over a completely randomized experiment can be achieved only when the degree of correlation, given by the ratio of the whole plot error variance to the sub-plot error variance, exceeds one. In many experimental situations, the completely randomized experiment will thus remain the best design option.

This result is confirmed by a factorial experiment in which the impact of the degree of correlation, the number of observations and the model on the features of the $\mathcal{D}$-optimal split-plot designs and on their efficiency relative to a completely randomized experiment was investigated. It turns out that the number of correlated observations in the optimal split-plot design increases with the degree of correlation and that the higher the correlation, the larger the efficiency gain of using a split-plot design instead of a completely randomized design. Also, the more complex the model, the larger the improvement in efficiency.

Finally, it is argued that the number of whole plots in a split-plot experiment should not be too small. Firstly, the design efficiency is often considerably improved when the number of whole plots is increased, all other things being equal. Next, the experimenter is better protected against systematic errors and possible trend effects when the factor levels are reset more often. In addition, more degrees of freedom are available for whole plot error estimation. Finally, even when the number of level changes is increased, split-plot designs are still easier to conduct than completely randomized designs.

Appendix. The construction algorithm

We denote the set of g candidate points and the set of design points by G and D respectively. Further, we denote by D_{jk} the set of design points belonging to the kth whole plot with whole plot factor level j and by λ_j the number of whole plots with whole plot factor level j. For simplicity, we will denote the information matrix $\mathbf{X'V^{-1}X}$ by $\mathbf{M}$. The singularity when $n < p$ is overcome by using $\mathbf{M} + \omega\mathbf{I}$ instead of $\mathbf{M}$ with ω a small positive number. We refer to the $\mathcal{D}$-criterion value of a design D as $\mathcal{D}$. The current optimal design is denoted by D^*, the corresponding criterion value is denoted by $\mathcal{D}^*$ and the whole plots of D^* are denoted by D^*_{jk}. Finally, we denote the number of tries by t and the number of the current try by t_c. The algorithm starts by specifying the set of grid points $G = \{1, 2, \ldots, g\}$ and proceeds as follows:

1. Set $\mathcal{D}^* = 0$ and $t_c = 1$.

2. Set $D = \emptyset$, $\mathbf{M} = \omega\mathbf{I}$, $\boldsymbol{\lambda} = \mathbf{0}$.

3. Generate starting design.
   - (a) Randomly choose r.
   - (b) Do r times:
     - i. Randomly choose $i \in G$ and denote its whole plot factor level by j.
     - ii. Randomly choose $k \in \{1, 2, \ldots, \lambda_j, \lambda_j + 1\}$.
     - iii. If $k > \lambda_j$, then set $\lambda_j = \lambda_j + 1$.
     - iv. $D_{jk} = D_{jk} \cup \{i\}$.
     - v. Update $\mathbf{M}$.
   - (c) Do $n - r$ times:
     - i. Determine $i \in G$ with largest prediction variance and denote its whole plot factor level by j.
     - ii. Randomly choose $k \in \{1, 2, \ldots, \lambda_j, \lambda_j + 1\}$.
     - iii. If $k > \lambda_j$, then set $\lambda_j = \lambda_j + 1$.
     - iv. $D_{jk} = D_{jk} \cup \{i\}$.
     - v. Update $\mathbf{M}$.

4. Compute $\mathbf{M}$ and $\mathcal{D}$. If $\mathcal{D} = 0$, then go to step 2, else continue.

5. Evaluate exchanges.

(a) Set $\delta = 1$.

(b) $\forall D_{jk}, \forall i \in D_{jk}, \forall l \in G$

    i. If the whole plot factor level of l equals j, then compute the effect $\delta_1 = \mathcal{D}'/\mathcal{D}$ of exchanging i by l in D_{jk}, else go to iii.

    ii. If $\delta_1 > \delta$, then $\delta = \delta_1$ and store information about the exchange.

    iii. Compute the effect $\delta_2 = \mathcal{D}'/\mathcal{D}$ of removing i from D_{jk} and assigning l to a new whole plot.

    iv. If $\delta_2 > \delta$, then $\delta = \delta_2$ and store information about the exchange.

    v. Compute the effect $\delta_3 = \mathcal{D}'/\mathcal{D}$ of removing i from D_{jk} and assigning l to an existing whole plot.

    vi. If $\delta_3 > \delta$, then $\delta = \delta_3$ and store information about the exchange.

6. If $\delta > 1$, then go to step 7, else go to step 8.

7. Carry out the best exchange, update $\mathbf{M}$, D and $\mathcal{D}$. Go to step 5.

8. If $\mathcal{D} \geq \mathcal{D}^*$, then $\mathcal{D}^* = \mathcal{D}$, $D^* = D$ and $D_{jk}^* = D_{jk}$.

9. If $t_c < t$, then $t_c = t_c + 1$ and go to step 2, else stop.

9

Two-Level Factorial and Fractional Factorial Designs

The design of two-level blocked and split-plot experiments has recently received a considerable amount of attention in the literature. In this chapter, we will give a brief overview of the recent developments. First, we introduce the 2^m and 2^{m-f} factorial designs and the concepts of resolution and aberration. Next, we discuss how the 2^m and 2^{m-f} factorial designs can be arranged in blocks and how they can be used in a split-plot experiment.

9.1 Introduction

Two-level factorial and fractional factorial designs are especially useful to identify the important variables. The designs are often used to fit first order models. The designs are described in many standard texts on experimental design and response surface methodology, for instance Khuri and Cornell (1987), Montgomery (1991), Atkinson and Donev (1992), Myers and Montgomery (1995) and Wu and Hamada (2000).

The 2^m and 2^{m-f} factorial designs, where m represents the number of experimental variables and f is the fraction of the 2^m design used, were already introduced in Section 1.5.1. The 2^3 factorial design is displayed in Table 1.5 and Figure 1.4. This design possesses $2^3 = 8$ runs and can be used to estimate the following model in three variables:

$$y = \beta_0 + \sum_{i=1}^{3} \beta_i x_i + \sum_{i=1}^{2} \sum_{j=i+1}^{3} \beta_{ij} x_i x_j + \beta_{123} x_1 x_2 x_3 + \varepsilon. \qquad (9.1)$$

Table 9.1: 2^3 factorial design.

Run	I	A	B	C	AB	AC	BC	ABC
1	+1	-1	-1	-1	+1	+1	+1	-1
2	+1	+1	-1	-1	-1	-1	+1	+1
3	+1	-1	+1	-1	-1	+1	-1	+1
4	+1	+1	+1	-1	+1	-1	-1	-1
5	+1	-1	-1	+1	+1	-1	-1	+1
6	+1	+1	-1	+1	-1	+1	-1	-1
7	+1	-1	+1	+1	-1	-1	+1	-1
8	+1	+1	+1	+1	+1	+1	+1	+1

As a result, the 2^3 factorial design can be used to estimate an intercept, the linear effects of the three factors, all two factor interactions and the three factor interaction. In total, eight different model parameters can be estimated. If we denote the three variables by A, B and C, the estimable effects can be represented by an I for the intercept, A, B and C for the linear effects, AB, AC and BC for the two factor interactions and ABC for the three factor interaction. The eight columns of the design matrix corresponding to model (9.1) are displayed in Table 9.1. In a similar fashion, 2^m factorial designs can be used to estimate a model containing 2^m unknown parameters: an intercept, m linear effects, $\binom{m}{2}$ two factor interaction effects, $\binom{m}{3}$ three factor interaction effects, ..., and one m-factor interaction effect.

In Section 1.5.1, it was shown that 2^m factorial designs are $\mathcal{D}$-optimal for estimating a model in the m variables containing the m linear terms and the interactions up to order m. The proof is also valid for all sub-models of this model, that is for all models that can be obtained by deleting one or more terms in the full model.

In practice, it is often infeasible to run the full 2^m factorial design. This is because the number of runs rapidly increases with the number of experimental variables m. Instead of running the full factorial design, only a fraction of it is used. Suppose, for example, that only four runs can be performed to estimate the effect of three experimental variables. The full 2^3 factorial design is then of course infeasible. One way to select four of its runs is to choose all the runs for which the ABC column has a +1. The resulting design is a half fraction of the 2^3 factorial design. It is therefore referred to as a $\frac{1}{2} \times 2^3 = 2^{3-1}$ fractional factorial design. The design is displayed in Table 9.2.

From the table, it can be seen that the columns corresponding to the intercept I and to the interaction ABC contain +1s. This is denoted by the

Table 9.2: 2^{3-1} fractional factorial design using I=ABC as the defining relation.

	I	A	B	C	AB	AC	BC	ABC
2	+1	+1	-1	-1	-1	-1	+1	+1
3	+1	-1	+1	-1	-1	+1	-1	+1
5	+1	-1	-1	+1	+1	-1	-1	+1
8	+1	+1	+1	+1	+1	+1	+1	+1

equality

$$I = ABC,$$

which is called the defining relation of the 2^{3-1} fractional factorial design. The interaction ABC is called the generator. Also the columns corresponding to the main effect of A and the two factor interaction BC are identical. Because of this equality, the linear effect of factor A and the BC interaction effect cannot be estimated independently from each other. The effects are therefore said to be confounded. A and BC are also called aliases. In a similar fashion, B and AC are confounded, as well as C and AB. Since, in the 2^{3-1} design, the main effects are confounded with two factor interactions, the design is said to be of resolution III. In general, a design is said to be of resolution R if no p-factor effect is aliased or confounded with another effect containing less than $R - p$ factors. A design is of resolution III if no main effects are confounded with other main effects, but main effects are aliased with two factor interactions, and two factor interactions may be aliased with each other. Similarly, a design is of resolution IV if no main effects are confounded with other main effects or with two factor interactions, but main effects are aliased with three factor interactions or two factor interactions are aliased with each other. The higher the resolution of a fractional factorial design, the better. This is because main effects and low order interactions are likely to be statistically significant, and, therefore, it is important to estimate them. On the contrary, high order interactions are mostly insignificant and their estimation is then unimportant. This explains why it is undesirable that main effects and low order interactions are confounded with each other and why it does not matter that much when high order interactions are aliased with main effects or low order interactions. In general, the resolution of a fractional factorial design is equal to the length of the shortest word —apart from I— in the defining relation. For the present example, the shortest word in the defining relation is ABC, as it is the only word. Its length is three, so that the corresponding design is of resolution III. In order to find out which effects are aliased with a given effect, we can simply multiply this effect with the defining relation. The resulting expression can be simplified using the facts that multiplying the column of an effect P with I, which is a column of ones, does not change the column of P and that multiplying a column by itself yields a column

Table 9.3: Constructing a 2^{3-1} fractional factorial design starting from a 2^2 factorial design.

Run	I	A	B	C AB
1	+1	-1	-1	+1
2	+1	+1	-1	-1
3	+1	-1	+1	-1
4	+1	+1	+1	+1

of ones and can be denoted by I. For instance, in order to find out what effect is confounded with the main effect A, we multiply I=ABC by A and obtain

$$IA = (ABC)A,$$
$$A = A^2BC,$$
$$A = IBC,$$
$$A = BC.$$

Above, we have shown how a 2^{3-1} fractional factorial design can be constructed from a full 2^3 factorial design. An alternative way to construct it is to start from the 2^2 design in the factor A and B and to assign the third factor C to the column corresponding to the interaction effect AB. It can be verified that this approach yields the 2^{3-1} design displayed in Table 9.3. In a similar fashion, half fractions of 2^m designs can be constructed. Again, we can start with the full factorial design in $m - 1$ factors and assign the mth factor to the $(m - 1)$-factor interaction effect. This method leads to the half fraction with the highest possible resolution.

For larger numbers of experimental variables, a quarter fraction or even a higher fraction of a 2^m design can be used. In order to construct a quarter fraction, we can start with the full factorial design in $m - 2$ variables. Now, two additional factors have to be assigned to an interaction effect of the first $m - 2$ variables. Suppose, for example, that a 2^{7-2} design is needed and that the factors are denoted by the letters A to G. One option is to assign the factors F and G to the columns corresponding to the interactions ABCD and ABCE respectively. The relations I=ABCDF and I=ABCEG are called the generating relations. The complete defining relation consists of the generating relations and the product I=(ABCDF)(ABCEG)=DEFG:

$$I=ABCDF=ABCEG=DEFG.$$

The shortest word in this defining relation is of length four so that the resulting fractional factorial design is of resolution IV. Note that the defining relation has one word of length four only. Another option would be to assign the factors F and G to the columns corresponding to the interactions

ABC and ADE respectively. The complete defining relation then becomes

$$I = ABCF = ADEG = BCDEFG.$$

The shortest word in this defining relation is also of length four. However, there are two words of length four. Therefore, the second option is inferior to the first one even though the resolution of the design also equals IV. The second design is said to have less aberration than the first one. If no design exists with less aberration than a given design, this design is said to have minimum aberration. For a formal definition of aberration, the word length pattern has to be defined. The word length pattern of a fractional factorial design is the vector

$$W = (A_3, A_4, \ldots, A_m),$$

where A_i is the number of words of length i in its defining relation. For the first option D_1 for the 2^{7-2} design, the word length pattern is

$$W(D_1) = (0, 1, 2, 0, 0),$$

whereas for the second option D_2, it is

$$W(D_2) = (0, 2, 0, 1, 0).$$

If we denote by r the smallest integer such that $A_r(D_1) \neq A_r(D_2)$, the design D_1 is said to have less aberration than D_2 if $A_r(D_1) < A_r(D_2)$.

The minimum aberration criterion is currently the most commonly used criterion to distinguish between fractional factorial designs of equal resolution. Chen, Sun and Wu (1993) proposed the number of clear effects as an additional criterion for design selection. An effect is clear if it is not confounded with a main effect or a two factor interaction. Minimum aberration 2^{m-f} fractional factorial designs for 8 to 128 runs are listed in Wu and Hamada (2000), who summarize the results of Chen et al. (1993), Chen (1992, 1998), Chen and Wu (1991), and Tang and Wu (1996).

It can be shown that 2^{m-f} fractional factorial designs are $\mathcal{D}$-optimal completely randomized designs for estimating main effects models with interactions provided the model does not contain aliased effects. It should be noted however that the $\mathcal{D}$-optimality criterion cannot distinguish between different 2^{m-f} fractional factorial designs.

9.2 Blocking 2^m and 2^{m-f} factorial designs

The 2^m and 2^{m-f} factorial designs discussed in the previous section can easily be arranged in blocks provided the block sizes are a power of two. The resulting blocked designs are $\mathcal{D}$-optimal for the random block effects

Table 9.4: 2^3 factorial design arranged in two blocks of size four.

Run	I	A	B	C	AB	AC	BC	ABC	Block
1	+1	-1	-1	-1	+1	+1	+1	-1	1
2	+1	+1	-1	-1	-1	-1	+1	+1	2
3	+1	-1	+1	-1	-1	+1	-1	+1	2
4	+1	+1	+1	-1	+1	-1	-1	-1	1
5	+1	-1	-1	+1	+1	-1	-1	+1	2
6	+1	+1	-1	+1	-1	+1	-1	-1	1
7	+1	-1	+1	+1	-1	-1	+1	-1	1
8	+1	+1	+1	+1	+1	+1	+1	+1	2

model and for the fixed block effects model. For a proof of the optimality, we refer the reader back to Section 4.3.1.

9.2.1 Blocking replicated 2^m or 2^{m-f} factorial designs

In some experimental situations, the blocks are large enough to contain a full 2^m factorial design or a 2^{m-f} fractional factorial design. In practice, the block sizes are often quite small, so that this approach is in most cases infeasible.

9.2.2 Blocking 2^m factorial designs

In order to arrange the eight runs of a 2^3 factorial experiment in two blocks of size four, the block effect, say B_1 can be confounded with the three-factor interaction ABC. In doing so, the four runs with a -1 in the ABC column in Table 9.1 are assigned to the first block, whereas the four runs with a +1 are assigned to the second block. The resulting blocked design is displayed in Table 9.4. The same design was encountered in Table 4.1 in Section 4.3.1.

Arranging the runs of the 2^3 design in $2^2 = 4$ blocks of size $2^1 = 2$ is more difficult. Now, two so-called independent blocking variables B_1 and B_2 are needed. One option is to confound them with the two factor interactions AB and BC. The resulting blocked design is displayed in Table 9.5. The runs for which B_1 and B_2 both equal -1 are assigned to block 1. The runs for which B_1 and B_2 equal -1 and +1 respectively are assigned to block 2, and so on. It can be verified from the table that B_1 can be interpreted as the difference between, on the one hand, blocks 1 and 2, and, on the other hand, blocks 3 and 4. Similarly, B_2 can be interpreted as the difference between, on the one hand, blocks 1 and 3, and, on the other hand, blocks 2 and 4. Finally, there is a third block effect that can be seen as the difference between the blocks 1 and 4, and the blocks 2 and 3. It can be verified from the table that this block effect is confounded with the interaction BC. This

Table 9.5: 2^3 factorial design arranged in four blocks of size two.

Run	I	A	B	C	B_1 AB	B_2 AC	BC	ABC	Block
1	+1	-1	-1	-1	+1	+1	+1	-1	4
2	+1	+1	-1	-1	-1	-1	+1	+1	1
3	+1	-1	+1	-1	-1	+1	-1	+1	2
4	+1	+1	+1	-1	+1	-1	-1	-1	3
5	+1	-1	-1	+1	+1	-1	-1	+1	3
6	+1	+1	-1	+1	-1	+1	-1	-1	2
7	+1	-1	+1	+1	-1	-1	+1	-1	1
8	+1	+1	+1	+1	+1	+1	+1	+1	4

is because $B_1B_2=(AB)(AC)=BC$. As a result, the blocking scheme used here confounds the three two factor interactions with the three blocking variables.

An alternative scheme to arrange the runs of the 2^3 factorial design is to choose $B_1 = ABC$ and $B_2 = AC$. This blocking scheme is however inferior because it confounds the main effect B with the third block effect. In fact, $B_1B_2=(ABC)(AC)=B$. Since it is vital that none of the main effects is confounded with a block effect, this alternative scheme is never used.

Now, if a general 2^m factorial design has to be blocked in $b = 2^q$ blocks of size $k = 2^{m-q}$, q independent block effects B_1, B_2, ..., B_q are needed. They are confounded with q interaction effects, say v_1, v_2, ..., v_q. The remaining block effects can then be obtained by multiplying the B_is:

$$B_1B_2 = v_1v_2, \ B_1B_3 = v_1v_3, \ \ldots \ , \ B_1B_2\ldots B_q = v_1v_2\ldots v_q.$$

The products v_1v_2, v_1v_3, ..., $v_1v_2\ldots v_q$ and v_1, v_2, ..., v_q form the socalled block defining contrast subgroup. For example, a 2^5 factorial design can be arranged in $2^3 = 8$ blocks of size $2^2 = 4$ using $B_1 = ACE$, $B_2 = BCE$ and $B_3 = ABCD$. The remaining block effects are confounded with $B_1B_2 = AB$, $B_1B_3 = BDE$, $B_2B_3 = ADE$, $B_1B_2B_3 = CD$, so that the block defining contrast subgroup of the design is

$$AB, CD, ACE, ADE, BCE, BDE, ABCD.$$

Another possible scheme is defined by using $B_1 = AB$, $B_2 = AC$ and $B_3 = DE$. The corresponding defining contrast subgroup is

$$AB, AC, BC, DE, ABDE, ACDE, BCDE.$$

This second scheme confounds four two factor interactions with the block effects, whereas the first scheme only confounds two two factor interactions. The first scheme therefore has less aberration than the first one and is considered superior. As a result, the minimum aberration criterion can also be used for constructing blocked experiments. Minimum aberration blocking schemes are tabulated in Sun, Wu and Chen (1997) and Wu and Hamada (2000).

9.2.3   Blocking 2^{m-f} fractional factorial designs

The choice of optimal blocking schemes for 2^{m-f} fractional factorial designs is more difficult than for 2^m factorial designs. This is because two defining contrast subgroups are needed: one for defining the blocking scheme and one for defining the fraction of the 2^m design. The former was called block defining contrast subgroup in the previous section. The latter defining contrast subgroup will be referred to as treatment defining contrast subgroup. The name treatment defining contrast subgroup is used because it determines what treatments, design points or factor level combinations will be used in the experiment. This contrast subgroup corresponds to the complete defining relation for 2^{m-f} designs introduced in Section 9.1. The two contrast subgroups jointly define the confounding pattern of the blocked 2^{m-f} design. Unfortunately, there is no natural extension to the concept of minimum aberration in this case. This is due to the presence of two defining contrast subgroups. Instead of looking for minimum aberration designs, the total number of clear effects is used as a criterion to distinguish between several design options. A clear effect is a main effect or a two factor interaction effect that is not confounded with any other main effect or two factor interaction, nor with any block effect.

Suppose, for instance, that a 2^{6-2} design has to be arranged in $2^2 = 4$ blocks of size $2^2 = 4$. In order to define the 2^{6-2} design, the following treatment defining contrast subgroup can be used:

$$I = ABCE = ABDF = CDEF. \tag{9.2}$$

In order to determine the blocking pattern, the following block defining contrast can be used:

$$B_1 = ACD, \ B_2 = BCD, \ \text{and} \ B_1 B_2 = AB. \tag{9.3}$$

By looking only at the block defining contrast subgroup, we might be tempted to say that the block effect B_1 is confounded with the three factor interaction ACD only. However, by multiplying ACD by the treatment defining subgroup contrast (9.2), we find that

$$ACD = BDE = BCF = AEF,$$

so that the block effect B_1 is also aliased with BDE, BCF and AEF. Similarly, it can be shown that B_2 is confounded with BCD, ADE, ACF and BEF, and that $B_1 B_2$ is confounded with AB, CE, DF and ABCDEF. As a result, the six main effects are not aliased with the block effects. However,

the two factor interactions AB, CE and DF are not clear. The 12 remaining
two factor interactions are aliased in pairs. This is no surprise as the words
in the treatment defining contrast subgroup contains three words of length
four. The aliasing pattern for the remaining two factor interactions is given
by

$$AC=BE=BCDF=ADEF,$$
$$AD=BF=BCDE=ACEF,$$
$$AE=BC=BDEF=ACDF,$$
$$AF=BD=BCEF=ACDE,$$
$$CD=EF=ABDE=ABCF,$$
$$CF=DE=ABEF=ABCD.$$

These expressions can be derived by multiplying each two factor interaction
by the treatment defining contrast subgroup (9.2). In a similar fashion, it
can be shown that all six main effects are clear:

$$A=BCE=BDF=ACDEF,$$
$$B=ACE=ADF=BCDEF,$$
$$C=ABE=DEF=ABCDF,$$
$$D=ABF=CEF=ABCDE,$$
$$E=ABC=CDF=ABDEF,$$
$$F=ABD=CDE=ABCEF.$$

The blocking of 2^{m-f} fractional factorial designs has received attention
by Sun et al. (1997) and Sitter, Chen and Feder (1997). These papers
contain an extensive list of blocked two-level fractional factorial designs. A
number of these designs are also tabulated in Wu and Hamada (2000). The
tables provide the reader with the best possible treatment defining contrast
subgroups and block defining contrast subgroups. In addition, the tables
contain information on the clear effects.

9.3 2^m and 2^{m-f} split-plot designs

In this section, a distinction is made between the hard-to-change and the
easy-to-change experimental variables. The former are called the whole plot
factors, while the latter are referred to as sub-plot factors. We will assume
in this section that there are m_w whole plot variables and $m_s = m -
m_w$ sub-plot variables. It is also assumed that each whole plot factor level
combination is visited once. As a result, the whole plot sizes are assumed
to be homogeneous and powers of two.

9.3.1 2^m split-plot designs

The design of a 2^m split-plot design is very easy. The experimental factors can be assigned to the columns of the 2^m design without taking into account the nature of the experimental variables. The resulting split-plot design has 2^{m_w} whole plots of size 2^{m_s} and is always crossed. In terms of $\mathcal{D}$-optimality, it is the best possible split-plot design with 2^{m_w} whole plots of size 2^{m_s} for estimating a linear model in m_w whole plot variables and m_s sub-plot variables with or without some or all of the interactions up to order m. This was shown in Section 6.5.1.

9.3.2 2^{m-f} split-plot designs

Like in the case of completely randomized 2^{m-f} designs, the minimum aberration criterion is mostly used to distinguish between 2^{m-f} split-plot designs. A two-level fractional factorial split-plot design is a split-plot design in which the whole plot variables are arranged as a fractional factorial design with fractional element f_w and the sub-plot variables are arranged as a fractional factorial design with fractional element f_s. This type of split-plot design is referred to as a $2^{(m_w+m_s)-(f_w+f_s)}$ fractional factorial split-plot design. It has $2^{m_w-f_w}$ whole plots of size $2^{m_s-f_s}$. The resulting design looks like an ordinary 2^{m-f} fractional factorial design, but the difference lies in the randomization structure.

In order to illustrate how a $2^{(m_w+m_s)-(f_w+f_s)}$ fractional factorial split-plot design can be constructed, consider an experiment with 16 runs to investigate the effect of seven experimental variables. Suppose that four of the variables are hard to change and denote them by A, B, C and D. Denote the remaining three variables by P, Q and R. Finally, suppose that eight whole plots of size two are available.

A simple way to construct a $2^{(m_w+m_s)-(f_w+f_s)}$ fractional factorial split-plot design is to use a $2^{m_w-f_w}$ fractional factorial design for the whole plot variables and a $2^{m_s-f_s}$ fractional factorial design for the sub-plot variables. The $2^{m_s-f_s}$ design for the sub-plot variables can then be run at each whole plot level, yielding a crossed split-plot design. First, a fractional factorial design is constructed for the whole plot factors. Since eight whole plot are available, a half fraction of a 2^4 factorial design is needed. One option is to use the defining relation I=ABCD, that is to assign the fourth whole plot variable D to the three factor interaction ABC. For the sub-plot variables a 2^{3-2} design can be used. The only defining relation that can be used for this design is I=PQ=PR. Combining the two defining relations yields the following defining contrast subgroup:

$$I=ABCD=PQ=PR=ABCDPQ=ABCDPR=QR=ABCDQR.$$

As this contrast subgroup contains words of length two, some main effects are confounded with other main effects and the resulting design has resolution II. The design is therefore very poor. Fortunately, a better split-plot design can be constructed by giving up the idea that the design in the split-plot variables should be a $2^{m_s - f_s}$ design. Instead of confounding the sub-plot factors with sub-plot factor effects only, we can also confound them with interactions involving whole plots.

In order to illustrate this approach, consider again the 16 run example. Again, we can start by constructing a half fraction of a 2^4 factorial design for the whole plot factors, for instance by using I=ABD as the defining relation. For the sub-plot factor levels, we can use the following assignments: Q=AP and R=CP. The resulting defining contrast subgroup is

$$I=ABD=APQ=CPR=ABCDPR=ACQR=BCDQR=BDPQ.$$

This subgroup does not contain words of length two, so that the new design has resolution III and is therefore better than the previous one. This example shows that allowing generators for the sub-plot design to contain whole plot variables gives the experimenter much more flexibility and makes it possible to construct better designs. The opposite, that is using sub-plot variables in the generators for the whole plot design, would destroy the split-plot nature of the experiment.

Bingham and Sitter (1999) extended the results of Huang, Chen and Voelkel (1998) and computed minimum aberration two-level fractional factorial designs. In their paper, they give a catalog of fractional factorial split-plot designs with 8 and 16 runs. The 16 run designs are displayed in an easier way in Bingham and Sitter (2001). For instance, the minimum aberration $2^{(4+3)-(1+2)}$ design uses I=ABCD as the defining relation for the whole plot design and I=ABPQ=ACPR for the sub-plot variables. The resulting defining contrast subgroup has words of four letters only:

$$I=ABCD=ABPQ=ACPR=ADQR=BCQR=BDPR=CDPQ.$$

As a result, the design, which is displayed in Table 9.6, is of resolution IV. This is substantially better than the designs constructed previously.

All minimum aberration designs listed in Huang, Chen and Voelkel (1998) and Bingham and Sitter (1999, 2001) possess the property that the levels of the sub-plot factors sum to zero within each whole plot. As was shown in Section 6.5.2, they are therefore $\mathcal{D}$-optimal.

Table 9.6: Minimum aberration $2^{(4+3)-(1+2)}$ split-plot design.

Run	A	B	C	P	D ABC	Q ABP	R ACP	Whole plot
1	-1	-1	-1	-1	-1	-1	-1	1
2	+1	-1	-1	-1	+1	+1	+1	2
3	-1	+1	-1	-1	+1	+1	-1	3
4	+1	+1	-1	-1	-1	-1	+1	4
5	-1	-1	+1	-1	+1	-1	+1	5
6	+1	-1	+1	-1	-1	+1	-1	6
7	-1	+1	+1	-1	-1	+1	+1	7
8	+1	+1	+1	-1	+1	-1	-1	8
9	-1	-1	-1	+1	-1	+1	+1	1
10	+1	-1	-1	+1	+1	-1	-1	2
11	-1	+1	-1	+1	+1	-1	+1	3
12	+1	+1	-1	+1	-1	+1	-1	4
13	-1	-1	+1	+1	+1	+1	-1	5
14	+1	-1	+1	+1	-1	-1	+1	6
15	-1	+1	+1	+1	-1	-1	-1	7
16	+1	+1	+1	+1	+1	+1	+1	8

9.4   Discussion

Two-level factorial and fractional factorial designs are extremely popular in industry. They are used in completely randomized experiments,
blocked experiments and split-plot experiments. In each of these experimental situations, they have excellent properties in terms of $\mathcal{D}$-optimality.
Unfortunately, two-level factorial and fractional factorial designs in many
cases do not provide the experimenter with a feasible design. This is, for
example, the case as soon as the number of runs available is not a power of
two. For blocked experiments, the applicability of the designs described in
this chapter is also limited because they require that the number of blocks
available and the block size are powers of two. Similarly, the two level factorial and fractional factorial designs described in the previous section can
only be used if the number of whole plots is a power of two, as well as the
whole plot size.

10
Summary and Future Research

In this book, the optimal design of two types of experiments with a restricted randomization has received elaborate attention. In the Chapters 4 and 5, we have concentrated on the design of blocked experiments. In the Chapters 6, 7 and 8, the focus was on split-plot experiments. In Chapter 9, the use of two-level factorial and fractional factorial designs in blocked and split-plot experiments was discussed.

In blocked experiments, the randomization is restricted because the experimental material is not homogeneous. This is often the case when not all the runs of the experiment can be carried out on the same day or when more than one batch of material is needed to perform all the experimental runs. The experimental runs that are conducted on the same day or using the same batch then compose a group of observations, which is referred to as a block. In this book, it was assumed that the blocks (days, batches) are randomly chosen from a population of blocks, and hence that the block effects are random instead of fixed. In a split-plot experiment, the randomization is restricted because the levels of some of the experimental factors are not independently reset for each run. Split-plotting is used especially when it is hard, time-consuming or expensive to change the levels of at least one experimental factor. Not resetting the factor levels for each run makes that split-plot experiments are considerably easier and cheaper to conduct than completely randomized experiments. In a split-plot experiment, the runs with the same level for the hard-to-change factors form a group of observations, which is called a whole plot.

It is clear that blocked and split-plot experiments have at least two common features: their runs are divided in groups and they possess two randomization procedures. In the first randomization procedure, the groups are randomized, while in the second, the runs are randomized within the groups. Therefore, both types of experiments belong to the class of bi-randomization or two-stratum experiments. The corresponding statistical model possesses two variance components, namely the group effect variance and the usual error variance. It assumes that observations from different groups are statistically independent, while the observations belonging to the same group have a compound symmetric error structure.

In spite of the broad use of blocked and split-plot experiments, their optimal design has received little attention in the literature. In this book, it was assumed that the interest was in estimating the factor effects on the response, not in the variance components. For each design problem considered, a point exchange algorithm for the construction of the best possible tailor-made experiment was described. The algorithms are generalizations of existing algorithms for computing $\mathcal{D}$-optimal completely randomized and blocked experiments. The compound symmetric error structure was exploited to speed up the algorithms.

In general, the optimal designs depend on the ratio of the group effect variance to the error variance. The larger this ratio, which is referred to as the degree of correlation, the more the observations within one group are correlated, and vice versa. However, it turns out that the optimal designs are robust against misspecification of the degree of correlation. Since it is usually unknown prior to the experiment, this robustness is extremely important from a practical point of view. In addition, we have found that the runs of the optimal designs for small degrees of correlation are in many cases identical to the runs of the optimal design for an experiment with uncorrelated observations. For larger degrees of correlation, this is no longer true. It will therefore not come as a surprise that substantial efficiency gains can be achieved by taking into account the correlation structure when designing a blocked or a split-plot experiment. The efficiency gain increases with the degree of correlation. It also turns out that assuming random block effects instead of fixed block effects leads to different designs. Finally, one of the most striking results of this book is that split-plot experiments are not only easier to carry out, but that they are often also statistically more efficient.

In addition to these computational results, the book contains a number of interesting theoretical results. Firstly, it is shown that orthogonal blocking is an optimal blocking strategy for a given set of experimental runs when the blocks are assumed to be fixed, independent of the block sizes. Secondly, it is shown that it is an optimal strategy as well when the block effects are random and the block size is homogeneous. In this case, ordinary and generalized least squares produce identical estimates of the intercept and the factor effects. The equivalence of ordinary and generalized least squares is no longer valid when the block size is heterogeneous. Thirdly, the optimality of some orthogonally blocked first order designs is established. Next, it is shown that the $\mathcal{D}$-optimal design of experiments with fixed blocks is a limiting case of the design of experiments with random block effects and it is proven that assigning the runs of a split-plot experiment to the whole plots so that the resulting experiment is crossed is the best possible assignment. Finally, the $\mathcal{D}$-optimality of 2^m and 2^{m-f} split-plot designs was established.

In this book, which is one of the first works on the optimal design of response surface experiments with correlated observations, the focus was on one specific correlation structure, namely the compound symmetric correlation structure. In other words, it was assumed that the extent to which observations are correlated is identical for every pair of observations within a group. However, it is not uncommon that experimental observations lying close to each other in time or space are more correlated than observations that lie far from each other. In contrast with the case of compound symmetry, the sequence of the experimental runs or the spatial location of the observations then plays a prominent role on the design efficiency. Of course, the exchange algorithms developed in this book will have to be modified to cope with this additional difficulty.

Another useful extension of the work contained within this book, is a generalization of two-stratum designs to multi-stratum designs. Trinca and Gilmour (2001) have developed an approach to design experiments of this kind. However, the choice of the experimental runs in their approach does not take into account the error structure of the experiment and the resulting design may therefore be inefficient. For the special case of a two-stratum experiment, this was illustrated in Chapter 6. A similar extension is the design of blocked experiments in which there is interaction between the factor effects and the block effects. Another extension, namely the design of blocked experiments in the presence of time trend effects, was studied by Tack and Vandebroek (2002).

Finally, it was assumed in this book that the interest was in estimating the factor effects, not in estimating the variance components. However, experimental situations exist in which the variance components are important. A detailed overview of the early work on the design of experiments for estimating variance components as well as the recent developments can be found in Khuri (2000). Khuri points out that the topic has received relatively little attention and concludes that many design problems in this area remain to be investigated. First of all, a consensus has to be reached on how to handle the design dependence on the variance components. Secondly, the simultaneous estimation of fixed effects and variance components deserves much more attention. Also, it is desirable to develop design and estimation procedures that are sequential in nature.

Bibliography

Abramowitz, M. and Stegun, I. A. (1970). *Handbook of Mathematical Functions*, New York: Dover Publications.

Anbari, F. T. and Lucas, J. M. (1994). Super-efficient designs: How to run your experiment for higher efficiency and lower cost, *ASQC Technical Conference Transactions*, pp. 852–863.

Anderson, V. L. and McLean, R. A. (1974). *Design of Experiments: A Realistic Approach*, New York: Marcel Dekker.

Atkins, J. E. (1994). *Constructing Optimal Designs in the Presence of Random Block Effects*, Ph.D. dissertation, University of California, Berkeley.

Atkins, J. E. and Cheng, C.-S. (1999). Optimal regression designs in the presence of random block effects, *Journal of Statistical Planning and Inference* **77**: 321–335.

Atkinson, A. C. and Cook, R. D. (1995). D-optimum designs for heteroscedastic linear models, *Journal of the American Statistical Association* **90**: 204–212.

Atkinson, A. C. and Donev, A. N. (1989). The construction of exact D-optimum experimental designs with application to blocking response surface designs, *Biometrika* **76**: 515–526.

Atkinson, A. C. and Donev, A. N. (1992). *Optimum Experimental Designs*, Oxford U.K.: Clarendon Press.

Berger, M. P. F. and Tan, F. E. S. (1998). Optimal designs for repeated measures experiments, *Kwantitatieve Methoden* **19**: 45–67.

Bingham, D. R. and Sitter, R. R. (2001). Design issues in fractional factorial split-plot designs, *Journal of Quality Technology* **33**: 2–15.

Bingham, D. and Sitter, R. R. (1999). Minimum-aberration two-level fractional factorial split-plot designs, *Technometrics* **41**: 62–70.

Bischoff, W. (1993). On D-optimal designs for linear models under correlated observations with an application to a linear model with multiple response, *Journal of Statistical Planning and Inference* **37**: 69–80.

Bisgaard, S. and Steinberg, D. M. (1997). The design and analysis of $2^{k-p} \times s$ prototype experiments, *Technometrics* **39**: 52–62.

Box, G. E. P. (1952). Multi-factor designs of first order, *Biometrika* **39**: 49–57.

Box, G. E. P. and Behnken, D. W. (1960). Some new three-level designs for the study of quantitative variables, *Technometrics* **2**: 455–475.

Box, G. E. P. and Draper, N. R. (1987). *Empirical Model-Building and Response Surfaces*, New York: Wiley.

Box, G. E. P. and Hunter, J. S. (1957). Multi-factor experimental designs for exploring response surfaces, *Annals of Mathematical Statistics* **28**: 195–241.

Box, G. E. P. and Jones, S. P. (1992). Split-plot designs for robust product experimentation, *Applied Statistics* **19**: 3–26.

Box, G. E. P. and Wilson, K. B. (1951). On the experimental attainment of optimum conditions, *Journal of the Royal Statistical Society, Ser. B* **13**: 1–45.

Box, M. J. and Draper, N. R. (1971). Factorial designs, the $|X'X|$ criterion, and some related matters, *Technometrics* **13**: 731–742. Correction, **14**, 511 (1972); **15**, 430 (1973).

Chasalow, S. D. (1992). *Exact Response Surface Designs with Random Block Effects*, Ph.D. dissertation, University of California, Berkeley.

Chasalow, S. D. and Brand, R. J. (1995). Generation of simplex lattice points, *Applied Statistics* **44**: 534–545.

Chen, J. (1992). Some results on 2^{n-k} fractional factorial designs and search for minimum aberration designs, *Annals of Statistics* **20**: 2124–2141.

Chen, J. (1998). Intelligent search for 2^{13-6} and 2^{14-7} minimum aberration designs, *Statistica Sinica* **8**: 1265–1270.

Chen, J., Sun, D. X. and Wu, C. F. J. (1993). A catalogue of two-level and three-level fractional factorial designs with small runs, *International Statistical Review* **61**: 131–145.

Chen, J. and Wu, C. F. J. (1991). Some results on s^{n-k} fractional factorial designs with minimum aberration or optimal moments, *Annals of Statistics* **19**: 1028–1041.

Cheng, C.-S. (1978). Optimality of certain asymmetrical experimental designs, *Annals of Statistics* **6**: 1239–1261.

Cheng, C.-S. (1995). Optimal regression designs under random block-effects models, *Statistica Sinica* **5**: 485–497.

Cliquet, S., Durier, C. and Kobilinsky, A. (1994). Principle of a fractional factorial design for qualitative and quantitative factors: Application to the production of bradyrhizobium japonicum in culture and inoculation media, *Agronomie* **14**: 569–587.

Cochran, W. G. and Cox, G. M. (1957). *Experimental Designs*, New York: Wiley.

Cook, R. D. and Nachtsheim, C. J. (1980). A comparison of algorithms for constructing exact D-optimal designs, *Technometrics* **22**: 315–324.

Cook, R. D. and Nachtsheim, C. J. (1989). Computer-aided blocking of factorial and response-surface designs, *Technometrics* **31**: 339–346.

Cornell, J. A. (1988). Analyzing data from mixture experiments containing process variables: A split-plot approach, *Journal of Quality Technology* **20**: 2–23.

Cornell, J. A. (1990). *Experiments with Mixtures: Designs, Models, and the Analysis of Mixture Data*, New York: Wiley.

Cornell, J. A. and Gorman, J. W. (1984). Fractional design plans for process variables in mixture experiments, *Journal of Quality Technology* **16**: 20–38.

Cox, D. R. (1958). *Planning of Experiments*, New York: Wiley.

Davies, H. M. (1945). *The Application of Variance Analysis to Some Problems of Petroleum Technology*, London: Publication of the Institute of Petroleum.

Davison, J. J. (1995). *Response Surface Designs and Analysis for Bi-Randomization Error Structures*, Ph.D. thesis, Virginia Polytechnic Institute and State University.

Donev, A. N. (1997). An algorithm for the construction of crossover trials, *Applied Statistics* **46**: 288–298.

Donev, A. N. (1998a). Construction of non-standard row-column designs, *in* R. Payne and P. Green (eds), *COMPSTAT 1998, Proceedings in Computational Statistics*, Harpenden: IACR-Rothamsted, pp. 275–280.

Donev, A. N. (1998b). Crossover designs with correlated observations, *Journal of Biopharmaceutical Statistics* **8**: 249–262.

Donev, A. N. and Atkinson, A. C. (1988). An adjustment algorithm for the construction of exact D-optimum experimental designs, *Technometrics* **30**: 429–433.

Draper, N. R. and Lin, D. K. J. (1990). Small response-surface designs, *Technometrics* **32**: 187–194.

Eccleston, J. A. and Chan, B. (1998). Design algorithms for correlated data, *in* R. Payne and P. Green (eds), *COMPSTAT 1998, Proceedings in Computational Statistics*, Harpenden: IACR-Rothamsted, pp. 41–52.

Farrell, R. H., Kiefer, J. and Walbran, A. (1967). Optimum multivariate designs, *Proceedings of the 5th Berkeley Symposium*, University of California Press, Berkeley, pp. 113–138.

Fedorov, V. V. (1972). *Theory of Optimal Experiments*, New York: Academic Press.

Fisher, R. A. (1935). *The Design of Experiments*, Edinburgh: Oliver & Boyd.

Fries, A. and Hunter, W. G. (1980). Minimum aberration 2^{k-p} designs, *Technometrics* **22**: 601–608.

Galil, Z. and Kiefer, J. (1980). Time- and space-saving computer methods, related to Mitchell's DETMAX, for finding D-optimum designs, *Technometrics* **21**: 301–313.

Ganju, J. (2000). On choosing between fixed and random block effects in some no-interaction models, *Journal of Statistical Planning and Inference* **90**: 323–334.

Ganju, J. and Lucas, J. M. (1997). Bias in test statistics when restrictions in randomization are caused by factors, *Communications in Statistics: Theory and Methods* **26**: 47–63.

Ganju, J. and Lucas, J. M. (1999). Detecting randomization restrictions caused by factors, *Journal of Statistical Planning and Inference* **81**: 129–140.

Gilmour, A. R., Thompson, R. and Cullis, B. R. (1995). An efficient algorithm for REML estimation in linear mixed models, *Biometrics* **51**: 1440–1450.

Gilmour, S. G. and Trinca, L. A. (2000). Some practical advice on polynomial regression analysis from blocked response surface designs, *Communications in Statistics: Theory and Methods* **29**: 2157–2180.

Goos, P., Tack, L. and Vandebroek, M. (2001). Optimal designs for variance function estimation using sample variances, *Journal of Statistical Planning and Inference* **92**: 233–252.

Goos, P. and Vandebroek, M. (1997). Semi-Bayesian D-optimal designs and estimation procedures for mean and variance functions, *Research Report 9745*, Katholieke Universiteit Leuven, Department of Applied Economics.

Goos, P. and Vandebroek, M. (2001a). D-optimal response surface designs in the presence of random block effects, *Computational Statistics and Data Analysis* **37**: 433–453.

Goos, P. and Vandebroek, M. (2001b). How to relax inconsistent constraints in a mixture experiment, *Chemometrics and Intelligent Laboratory Systems* **55**: 147–149.

Goos, P. and Vandebroek, M. (2001c). Optimal split-plot designs, *Journal of Quality Technology* **33**: 436–450.

Guest, P. G. (1958). The spacing of observations in polynomial regression, *Annals of Mathematical Statistics* **29**: 294–299.

Haines, L. M. (1987). The application of the annealing algorithm to the construction of exact optimal designs for linear-regression models, *Technometrics* **29**: 439–447.

Hartley, H. O. (1959). Smallest composite designs for quadratic response surfaces, *Biometrics* **15**: 611–624.

Harville, D. A. (1997). *Matrix Algebra from a Statistician's Perspective*, New York: Springer.

Hoel, P. G. (1958). Efficiency problems in polynomial estimation, *Annals of Mathematical Statistics* **29**: 1134–1145.

Hoke, A. T. (1974). Economical second-order designs based on irregular fractions of the 3^n factorial, *Technometrics* **16**: 375–384.

Huang, P., Chen, D. and Voelkel, J. (1998). Minimum-aberration two-level split-plot designs, *Technometrics* **40**: 314–326.

John, J. A. and Williams, E. R. (1995). *Cyclic and Computer Generated Designs*, London: Chapman & Hall.

Johnson, M. E. and Nachtsheim, C. J. (1983). Some guidelines for constructing exact D-optimal designs on convex design spaces, *Technometrics* **25**: 271–277.

Kacker, R. N. and Harville, D. A. (1984). Approximations for standard errors of estimators of fixed and random effects in mixed linear models, *Journal of the American Statistical Society* **79**: 853–862.

Karlin, S. and Studden, W. J. (1966). Optimal experimental designs, *Annals of Mathematical Statistics* **37**: 783–815.

Kempthorne, O. (1952). *The Design and Analysis of Experiments*, New York: Wiley.

Kenward, M. G. and Roger, J. H. (1997). Small sample inference for fixed effects from restricted maximum likelihood, *Biometrics* **53**: 983–997.

Khuri, A. I. (1992). Response surface models with random block effects, *Technometrics* **34**: 26–37.

Khuri, A. I. (1994). Effect of blocking on the estimation of a response surface, *Journal of Applied Statistics* **21**: 305–316.

Khuri, A. I. (2000). Designs for variance components estimation: Past and present, *International Statistical Review* **68**: 311–322.

Khuri, A. I. and Cornell, J. A. (1987). *Response Surfaces: Designs and Analyses*, New York: Marcel Dekker.

Kiefer, J. (1959). Optimal experimental designs (with discussion), *Journal of the Royal Statistical Society, Ser. B* **21**: 272–319.

Kiefer, J. (1961). Optimum design in regression problems II, *Annals of Mathematical Statistics* **32**: 298–325.

Kiefer, J. (1971). The role of symmetry and approximation in exact design optimality, *in* S. S. Gupta and J. Yachel (eds), *Statistical Design Theory and Related Topics*, New York: Academic Press, pp. 109–118.

Kiefer, J. (1975). Optimal design: variation in structure and performance under change of criterion, *Biometrika* **62**: 277–288.

Kowalski, S. M., Cornell, J. A. and Vining, G. G. (2002). Split-plot designs and estimation methods for mixture experiments with process variables, *Technometrics* **44**: 72–79.

Kowalski, S. M. and Vining, G. G. (2001). Split-plot experimentation for process and quality improvement, *in* H. Lenz (ed.), *Frontiers in Statistical Quality Control 6*, Heidelberg: Springer-Verlag, pp. 335–350.

Kunert, J. (1991). Cross-over designs for two treatments and correlated errors, *Biometrika* **78**: 315–324.

Kurotschka, V. G. (1981). A general approach to optimum design of experiments with qualitative and quantitative factors, *in* J. Ghosh and J. Roy (eds), *Proceedings of the Indian Statistical Institute Golden Jubilee International Conference on Statistics: Applications and New Directions*, Indian Statistical Institute, Calcutta, pp. 353–368.

Letsinger, J. D., Myers, R. H. and Lentner, M. (1996). Response surface methods for bi-randomization structures, *Journal of Quality Technology* **28**: 381–397.

Lucas, J. M. (1974). Optimum composite designs, *Technometrics* **16**: 561–567.

Lucas, J. M. (1976). Which response surface design is best, *Technometrics* **18**: 411–417.

Lucas, J. M. (1977). Design efficiencies for varying numbers of center points, *Biometrika* **64**: 145–147.

Lucas, J. M. and Ju, H. L. (2002). L^k factorial experiments with hard-to-change and easy-to-change factors. To appear in *Journal of Quality Technology*.

Martin, R. J., Eccleston, J. A. and Jones, G. (1998). Some results on multilevel factorial designs with dependent observations, *Journal of Statistical Planning and Inference* **73**: 91–111.

Martin, R. J., Jones, G. and Eccleston, J. A. (1998). Some results on two-level factorial designs with dependent observations, *Journal of Statistical Planning and Inference* **66**: 363–384.

Mays, D. P. (1999). Optimal central composite designs in the presence of dispersion effects, *Journal of Quality Technology* **31**: 398–407.

Mays, D. P. and Easter, S. M. (1997). Optimal response surface designs in the presence of dispersion effects, *Journal of Quality Technology* **29**: 59–70.

Mee, R. (2001). Noncentral composite designs, *Technometrics* **43**: 34–43.

Meyer, R. K. and Nachtsheim, C. J. (1988). Simulated annealing in the construction of exact optimal design of experiments, *American Journal of Mathematical and Management Science* **8**: 329–359.

Meyer, R. K. and Nachtsheim, C. J. (1995). The coordinate-exchange algorithm for constructing exact optimal experimental designs, *Technometrics* **37**: 60–69.

Mitchell, T. J. (1974a). An algorithm for the construction of D-optimal experimental designs, *Technometrics* **16**: 203–210.

Mitchell, T. J. (1974b). Computer construction of D-optimal first-order designs, *Technometrics* **16**: 211–220.

Montgomery, D. C. (1991). *Design and Analysis of Experiments*, New York: Wiley.

Müller, W. G. (1998). *Collecting Spatial Data*, New York: Physica-Verlag.

Myers, R. H. and Montgomery, D. C. (1995). *Response Surface Methodology: Process and Product Optimization using Designed Experiments*, New York: Wiley.

Näther, W. (1985). *Effective Observation of Random Fields*, Leipzig: Teubner Verlag.

Nelson, L. S. (1985). What do low F ratios tell you?, *Journal of Quality Technology* **17**: 237–238.

Neter, J., Kutner, M. H., Nachtsheim, C. J. and Wasserman, W. (1996). *Applied Linear Statistical Models*, London: Irwin.

Nguyen, N.-K. and Miller, A. J. (1992). A review of some exchange algorithms for constructing discrete d-optimal designs, *Computational Statistics and Data Analysis* **14**: 489–498.

Piepel, G. F. (1988). Programs for generating extreme vertices and centroids of linearly constrained experimental regions, *Journal of Quality Technology* **20**: 125–139.

Plackett, R. L. and Burman, J. P. (1946). The design of optimum multifactorial experiments, *Biometrika* **33**: 305–325.

Pukelsheim, F. (1993). *Optimal Design of Experiments*, New York: Wiley.

Roquemore, K. G. (1976). Hybrid designs for quadratic response surfaces, *Technometrics* **18**: 419–423.

Scheffé, H. (1958). Experiments with mixtures, *Journal of the Royal Statistical Society, Ser. B* **20**: 344–360.

Schoenberg, I. J. (1959). On the maxima of certain Hankel determinants and the zeros of the classical orthogonal polynomials, *Indagationes Mathematicae* **21**: 282–290.

Shah, K. R. and Sinha, B. K. (1989). *Theory of Optimal Designs*, New York: Springer-Verlag.

Silvey, S. D. (1980). *Optimal Design*, London: Chapman and Hall.

Sitter, R. R., Chen, J. and Feder, M. (1997). Fractional resolution and minimum aberration in blocked 2^{n-p} designs, *Technometrics* **39**: 382–390.

Smith, K. (1918). On the standard deviations of adjusted and interpolated values of an observed polynomial function and its constraints and the guidance they give towards a proper distribution of observations, *Biometrika* **12**: 1–85.

Sun, D. X., Wu, C. F. J. and Chen, Y. (1997). Optimal blocking schemes for 2^n and 2^{n-p} designs, *Technometrics* **39**: 298–307.

Tack, L., Goos, P. and Vandebroek, M. (2002). Efficient Bayesian designs under heteroscedasticity, *Journal of Statistical Planning and Inference* **104**: 391–405.

Tack, L. and Vandebroek, M. (2002). Trend-resistant and cost-efficient block designs with fixed or random block effects. To appear in *Journal of Quality Technology*.

Taguchi, G. (1989). *Introduction to Quality Engineering: Designing Quality into Products and Processes*, Tokyo: Asian Productivity Organization.

Tang, B. and Wu, C. F. J. (1996). Characterization of the minimum aberration 2^{n-k} designs in terms of their complementary designs, *Annals of Statistics* **24**: 2549–2559.

Trinca, L. A. and Gilmour, S. G. (1999). Difference variance dispersion graphs for comparing response surface designs with applications in food industry, *Applied Statistics* **48**: 441–455.

Trinca, L. A. and Gilmour, S. G. (2000). An algorithm for arranging response surface designs in small blocks, *Computational Statistics and Data Analysis* **33**: 25–43.

Trinca, L. A. and Gilmour, S. G. (2001). Multi-stratum response surface designs, *Technometrics* **43**: 25–33.

Vining, G. G. and Myers, R. H. (1990). Combining Taguchi and response surface philosophies: A dual response approach, *Journal of Quality Technology* **22**: 38–45.

Vining, G. and Schaub, D. (1996). Experimental design for estimating both mean and variance functions, *Journal of Quality Technology* **28**: 135–147.

Welch, W. J. (1982). Branch-and-bound search for experimental designs based on D-optimality and other criteria, *Technometrics* **24**: 41–48.

Welch, W. J. (1984). Computer-aided design of experiments for response estimation, *Technometrics* **26**: 217–224.

Wu, C. F. J. and Hamada, M. (2000). *Experiments: Planning, Analysis, and Parameter Design Optimization*, New York: Wiley.

Wynn, H. P. (1972). Results in the theory and construction of D-optimum experimental designs, *Journal of the Royal Statistical Society, Ser. B* **34**: 133–147.

Index

Lecture Notes in Statistics

For information about Volumes 1 to 110, please contact Springer-Verlag

111: Leon Willenborg and Ton de Waal, Statistical Disclosure Control in Practice. xiv, 152 pp., 1996.

112: Doug Fischer, Hans-J. Lenz (Editors), Learning from Data. xii, 450 pp., 1996.

113: Rainer Schwabe, Optimum Designs for Multi-Factor Models. viii, 124 pp., 1996.

114: C.C. Heyde, Yu. V. Prohorov, R. Pyke, and S. T. Rachev (Editors), Athens Conference on Applied Probability and Time Series Analysis, Volume I: Applied Probability In Honor of J.M. Gani. viii, 424 pp., 1996.

115: P.M. Robinson and M. Rosenblatt (Editors), Athens Conference on Applied Probability and Time Series Analysis, Volume II: Time Series Analysis In Memory of E.J. Hannan. viii, 448 pp., 1996.

116: Genshiro Kitagawa and Will Gersch, Smoothness Priors Analysis of Time Series. x, 261 pp., 1996.

117: Paul Glasserman, Karl Sigman, and David D. Yao (Editors), Stochastic Networks. xii, 298 pp., 1996.

118: Radford M. Neal, Bayesian Learning for Neural Networks. xv, 183 pp., 1996.

119: Masanao Aoki and Arthur M. Havenner, Applications of Computer Aided Time Series Modeling. ix, 329 pp., 1997.

120: Maia Berkane, Latent Variable Modeling and Applications to Causality. vi, 288 pp., 1997.

121: Constantine Gatsonis, James S. Hodges, Robert E. Kass, Robert McCulloch, Peter Rossi, and Nozer D. Singpurwalla (Editors), Case Studies in Bayesian Statistics, Volume III. xvi, 487 pp., 1997.

122: Timothy G. Gregoire, David R. Brillinger, Peter J. Diggle, Estelle Russek-Cohen, William G. Warren, and Russell D. Wolfinger (Editors), Modeling Longitudinal and Spatially Correlated Data. x, 402 pp., 1997.

123: D. Y. Lin and T. R. Fleming (Editors), Proceedings of the First Seattle Symposium in Biostatistics: Survival Analysis. xiii, 308 pp., 1997.

124: Christine H. Müller, Robust Planning and Analysis of Experiments. x, 234 pp., 1997.

125: Valerii V. Fedorov and Peter Hackl, Model-Oriented Design of Experiments. viii, 117 pp., 1997.

126: Geert Verbeke and Geert Molenberghs, Linear Mixed Models in Practice: A SAS-Oriented Approach. xiii, 306 pp., 1997.

127: Harald Niederreiter, Peter Hellekalek, Gerhard Larcher, and Peter Zinterhof (Editors), Monte Carlo and Quasi-Monte Carlo Methods 1996. xii, 448 pp., 1997.

128: L. Accardi and C.C. Heyde (Editors), Probability Towards 2000. x, 356 pp., 1998.

129: Wolfgang Härdle, Gerard Kerkyacharian, Dominique Picard, and Alexander Tsybakov, Wavelets, Approximation, and Statistical Applications. xvi, 265 pp., 1998.

130: Bo-Cheng Wei, Exponential Family Nonlinear Models. ix, 240 pp., 1998.

131: Joel L. Horowitz, Semiparametric Methods in Econometrics. ix, 204 pp., 1998.

132: Douglas Nychka, Walter W. Piegorsch, and Lawrence H. Cox (Editors), Case Studies in Environmental Statistics. viii, 200 pp., 1998.

133: Dipak Dey, Peter Müller, and Debajyoti Sinha (Editors), Practical Nonparametric and Semiparametric Bayesian Statistics. xv, 408 pp., 1998.

134: Yu. A. Kutoyants, Statistical Inference For Spatial Poisson Processes. vii, 284 pp., 1998.

135: Christian P. Robert, Discretization and MCMC Convergence Assessment. x, 192 pp., 1998.

136: Gregory C. Reinsel, Raja P. Velu, Multivariate Reduced-Rank Regression. xiii, 272 pp., 1998.

137: V. Seshadri, The Inverse Gaussian Distribution: Statistical Theory and Applications. xii, 360 pp., 1998.

138: Peter Hellekalek and Gerhard Larcher (Editors), Random and Quasi-Random Point Sets. xi, 352 pp., 1998.

139: Roger B. Nelsen, An Introduction to Copulas. xi, 232 pp., 1999.

140: Constantine Gatsonis, Robert E. Kass, Bradley Carlin, Alicia Carriquiry, Andrew Gelman, Isabella Verdinelli, and Mike West (Editors), Case Studies in Bayesian Statistics, Volume IV. xvi, 456 pp., 1999.

141: Peter Müller and Brani Vidakovic (Editors), Bayesian Inference in Wavelet Based Models. xiii, 394 pp., 1999.

142: György Terdik, Bilinear Stochastic Models and Related Problems of Nonlinear Time Series Analysis: A Frequency Domain Approach. xi, 258 pp., 1999.

143: Russell Barton, Graphical Methods for the Design of Experiments. x, 208 pp., 1999.

144: L. Mark Berliner, Douglas Nychka, and Timothy Hoar (Editors), Case Studies in Statistics and the Atmospheric Sciences. x, 208 pp., 2000.

145: James H. Matis and Thomas R. Kiffe, Stochastic Population Models. viii, 220 pp., 2000.

146: Wim Schoutens, Stochastic Processes and Orthogonal Polynomials. xiv, 163 pp., 2000.

147: Jürgen Franke, Wolfgang Härdle, and Gerhard Stahl, Measuring Risk in Complex Stochastic Systems. xvi, 272 pp., 2000.

148: S.E. Ahmed and Nancy Reid, Empirical Bayes and Likelihood Inference. x, 200 pp., 2000.

149: D. Bosq, Linear Processes in Function Spaces: Theory and Applications. xv, 296 pp., 2000.

150: Tadeusz Caliński and Sanpei Kageyama, Block Designs: A Randomization Approach, Volume I: Analysis. ix, 313 pp., 2000.

151: Håkan Andersson and Tom Britton, Stochastic Epidemic Models and Their Statistical Analysis. ix, 152 pp., 2000.

152: David Ríos Insua and Fabrizio Ruggeri, Robust Bayesian Analysis. xiii, 435 pp., 2000.

153: Parimal Mukhopadhyay, Topics in Survey Sampling. x, 303 pp., 2000.

154: Regina Kaiser and Agustin Maravall, Measuring Business Cycles in Economic Time Series. vi, 190 pp., 2000.

155: Leon Willenborg and Ton de Waal, Elements of Statistical Disclosure Control. xvii, 289 pp., 2000.

156: Gordon Willmot and X. Sheldon Lin, Lundberg Approximations for Compound Distributions with Insurance Applications. xi, 272 pp., 2000.

157: Anne Boomsma, Marijtje A.J. van Duijn, and Tom A.B. Snijders (Editors), Essays on Item Response Theory. xv, 448 pp., 2000.

158: Dominique Ladiray and Benoît Quenneville, Seasonal Adjustment with the X-11 Method. xxii, 220 pp., 2001.

159: Marc Moore (Editor), Spatial Statistics: Methodological Aspects and Some Applications. xvi, 282 pp., 2001.

160: Tomasz Rychlik, Projecting Statistical Functionals. viii, 184 pp., 2001.

161: Maarten Jansen, Noise Reduction by Wavelet Thresholding. xxii, 224 pp., 2001.

162: Constantine Gatsonis, Bradley Carlin, Alicia Carriquiry, Andrew Gelman, Robert E. Kass Isabella Verdinelli, and Mike West (Editors), Case Studies in Bayesian Statistics, Volume V. xiv, 448 pp., 2001.

163: Erkki P. Liski, Nripes K. Mandal, Kirti R. Shah, and Bikas K. Sinha, Topics in Optimal Design. xii, 164 pp., 2002.

164: Peter Goos, The Optimal Design of Blocked and Split-Plot Experiments. xiv, 244 pp., 2002.